Factoring Ethics in Technology, Policy Making, Regulation and AI

Edited by Ali G. Hessami and Patricia Shaw

Published in London, United Kingdom

IntechOpen

Supporting open minds since 2005

Factoring Ethics in Technology, Policy Making, Regulation and AI
http://dx.doi.org/10.5772/intechopen.92952
Edited by Ali G. Hessami and Patricia Shaw

Contributors
Clement Longondjo Etambakonga, Claudio Alvarez, Gustavo Zurita, Beatriz Hasbun, Sergio Peñafiel, Alvaro Pezoa, Ram Mohan Rao P, AP Siva Kumar, S. Murali Krishna, David Fourie, Cornel Malan, M. Victoria Lopez-Perez, Lourdes Arco-Castro, Mauricio Florez-Parra, Sara Rodriguez-Gomez, Dave Lewis, David Filip, Zvikomborero Murahwi, Ali G. Hessami, Patricia Shaw, Harshvardhan J. Pandit

First published in London, United Kingdom, 2021 by IntechOpen
IntechOpen is the global imprint of INTECHOPEN LIMITED, registered in England and Wales, registration number: 11086078, 5 Princes Gate Court, London, SW7 2QJ, United Kingdom
Printed in Croatia

British Library Cataloguing-in-Publication Data
A catalogue record for this book is available from the British Library

Additional hard and PDF copies can be obtained from orders@intechopen.com

Factoring Ethics in Technology, Policy Making, Regulation and AI
Edited by Ali G. Hessami and Patricia Shaw
p. cm.
Print ISBN 978-1-83969-160-7
Online ISBN 978-1-83969-161-4
eBook (PDF) ISBN 978-1-83969-162-1

For EU product safety concerns:
IN TECH d.o.o., Prolaz Marije Krucifikse Kozulić 3, 51000 Rijeka, Croatia, info@intechopen.com or visit our website at intechopen.com.

Meet the editors

Ali Hessami is Director of R&D and Innovation at Vega Systems. He is an expert in systems assurance and safety, security, sustainability, and knowledge assessment/management methodologies and has a background in the design and development of advanced control systems for business and safety-critical industrial applications. He chairs the Institute of Electrical and Electronics Engineers (IEEE) 7000 Technology Ethics standard and is the vice-chair of the Ethics Certification ECPAIS Programme at IEEE. Dr. Hessami is a visiting professor at London City University's Centre for Systems and Control and at Beijing Jiaotong University School of Electronics and Information Engineering. He is also a fellow of the Royal Society of Arts (FRSA), a fellow of the UK Institution of Engineering and Technology (IET), and a Life Senior Member of IEEE.

With twenty years of experience as a solicitor in technology, regulatory/government affairs, and risk management, Patricia Shaw advises on artificial intelligence (AI) and data ethics, policy, and governance, including ethics advisory boards. She is a public speaker and author, with contributions in *The AI Book*,[1] *The Law of AI*,[2] and *Responsible AI: A Global Policy Framework Update 2021*.[3] She also chairs the UK Society for Computers and Law, sits on the board of and is vice-chair of the AI Committee for iTechLaw, and is an advocate for responsible AI through her work with ForHumanity, Women Leading in AI, and the Institute of Electrical and Electronics Engineers (IEEE).

1 https://fintechcircle.com/ai-book/

2 Sweet and Maxwell's The Law of Artificial Intelligence: https://www.sweetandmaxwell.co.uk/Product/Information-Technology-Law/The-Law-of-Artificial-Intelligence/Hardback/42675054

3 International Technology Law Association, Responsible AI: A Global Policy Framework Update 2021: https://www.itechlaw.org/ResponsibleAI2021

Contents

Preface

A few millennia of recorded civilisation has now reached a stage that respect for human rights and values has emerged as a major global challenge largely driven by the advent and widespread deployment of Autonomous Decision Making and Algorithmic Learning Systems (ADM/ALS). The potential threat to human agency, rights, values, and freedoms arising from human-made artefacts and Artificial Intelligence (AI) is largely behind the recent revival of interest in ethics. This is analogous to the mind-body dualism of Rene Descartes that placed mankind on top of the natural order, and nature as subservient to man's intellect and to be exploited. Spinoza, a prominent philosopher of the Age of Enlightenment challenged Cartesian dualism by pointing out that while humans and nature are different, they represent aspects of a unified grand reality. The current trends in technology ethics resemble the rebirth of a new dualism, humans versus the artefacts of their mind and endeavours, autonomous intelligent machines.

This book informs decision-makers and practitioners about best practices in the emerging field of technology and AI ethics.

The chapters fall into three categories to guide the readers to gain insight from generic fundamentals to discipline-specific case studies and state of practice in technology ethics. The three broad categories are:

1. Environmental, Social and Governance

2. Ethical Best Practice

3. Ethical Education

For sharing their knowledge, their professional stance, hard efforts, and diligence in reviewing and implementing the proposed enhancements to their manuscript, we would like to thank all authors who have contributed to this book. We have also incorporated reviewers' recommendations and the book is much improved as a consequence.

Special thanks goes to Author Service Manager Sara Gojević-Zrnić at IntechOpen for her prudence, dedicated support in the submission and reviewing process, and her suggestions for further improvement. Finally, all thanks to IntechOpen for publishing this book.

Ali G. Hessami, Ph.D., BSc (Hons), EurIng, CEng, FIET, SMIEEE
Professor,
Director of R&D and Innovation Vega Systems,
London, United Kingdom

Patricia Shaw LLB(Hons), LLM, FRSA
CEO and Founder,
Beyond Reach – a Tech Ethics (Governance, Policy and Legal) Consultancy,
Greater Nottingham, United Kingdom

Section 1

Environmental, Social and Governance

Chapter 1

Introductory Chapter: AI's Very Unlevel Playing Field

Ali Hessami and Patricia Shaw

1. Introduction

There are many great initiatives happening in the space of AI and data ethics. There are a variety of high-level principles, process and procedural standards, risk and impact assessments, certification, and audit all in the making. However, to move from voluntary adoption and inconsistent application of such great works, will require robust policy and ultimately law.

Calling for the need for the regulation of AI, is not innovation stifling. Instead it has the potential to birth an industry, create a level of reliability and safety for people and planet that has not been previously secured, to embed human dignity, human flourishing, human autonomy, freedom of choice and non-discrimination into AI design, development, deployment, monitoring and decommissioning. AI is formed in a lifecycle but implemented and operationalised in a diverse ecosystem which contains contrasting and competing interests, where context truly matters. Therefore, no regulatory response should avoid this complexity but tackle the challenge head on in a manner which enables it to flex to a constantly changing technological environments and be sufficiently agile and adaptable to be future-proof. It will require learning lessons from the history of AI products and services, and a deep dive into the possibilities of existing and emerging technologies that lie before us.

We need to understand both the risk and the likelihood of the risk impact occurring in the short, medium and long term, and how this risk and its impact changes from context to context, country to country, culture to culture. Diversity, equity and inclusion matter.

We must evaluate our infrastructure, our governance frameworks, and design a regulatory response that is fit for purpose both from the top down and bottom up. Better business, better outcomes, better society, can all be born out of greater stakeholder engagement and participatory governance in AI. Recurrent and dynamic feedback should be our weapon of choice to head not just legal but also ethical risk off at the pass, to abate biased or unfair and exclusionary outcomes, technologically disguised anti-competitive behaviours, unintentional consumer and citizen harm, indirect and inadvertent discrimination, and ultimately unconscious human rights impacts and infringements.

We cannot preserve the status quo otherwise we will simply sleepwalk into making the same mistakes of the past, embedding historic and systemic attributes and risks. Compliance on a mere voluntary or "soft law" basis will not simply cut the mustard. Neither will a siloed jurisdictional approach to AI regulation. Cohesion, cooperation, collaboration will be key for any new regulatory system which seeks to transcend regulatory arenas and cross national borders.

The development of a globalised AI ecosystem sets the requirement for an umbrella international regulatory response. The global–local dichotomy and paradox can no longer be ignored. Utilising generic data and AI tools to apply generalisations from one jurisdiction to the next must be called out for what they are "irresponsible AI" based on a lucky guessbet. This is where the importance of localism, culture and context will come into their own. An international regulatory response will be most effective where it brings parties together with a consistent nomenclature with regularisation tools such as standards, certification and audit which must be built with globally diverse and inclusive actors. It will not be at its best by effectuating what can only be described as AI ethics colonialism, seeking to apply one set of ethics to all contexts. This is why application and enforcement should be (and ought always be) best left in the hands of the relevant jurisdiction(s). This way AI ethics can be both contextually. Culturally and equitably applied.

2. The case for ethics

> *"[There is a] need for Governments, the private sector, international organizations, civil society, the technical and academic communities and all relevant stakeholders to be cognizant of the impact, opportunities and challenges of rapid technological change on the promotion and protection of human rights, as well as of its potential to facilitate efforts, to accelerate human progress and to promote and protect human rights and fundamental freedoms" [1].*

Ethics is a forerunner to legislation. It is the ethical dilemmas that we face as a society that prompt the need for new law, where existing laws do not or cannot fill the gaps. Law provides certainty and resolve to the social problem faced. At some point in every generation, a society has to decide what is acceptable or not acceptable behaviour or outcomes. This generation no less than any generation that has gone before it. Simply this time it concerns AI, or more pertinently the use of big data and algorithmic intelligent and (semi-) autonomous systems, the control (or not) the designers, developers and deployers, and those that monitor their performance, have over and concerning the outcomes.

The AI ecosystem and supply chain are complex which makes legislating it and ensuring that it is futureproof too so very tricky.

AI ethics itself has so far proven popular. It has certainly raised the issue of "trustworthy" [2]. AI not only at a national but international scale. The challenge is, has AI ethics alone really changed anything at all! Being a leader and a responsible AI advocate can be a real competitive advantage, but this is where operationalising AI ethics moves the goalposts from Advocate to Actor to Ambassador, where AI ethics and the governance that operationalises it can become a real innovation enabler.

Change must occur. But it can only do so if governments and businesses are willing to take the first steps to learn how to operationalise AI ethics, and embed agile and dynamic governance which works in harmony with its stakeholders.

AI ethics is not merely about securing privacy (or more pertinently data privacy) for end users. Although that is a step in the right direction. It is about creating an equitable digital society whereby human and organisational, socio- and technical tools work towards trustworthy human oversight, informed human agency, and good exercise of human autonomy. A move away from bias, underrepresented people groups and lack of diversity, towards fairer and non-discriminatory outcomes; allowing for appropriate process and procedural transparency as well as decision transparency, not just transparency of data, models and code, to ensure the necessary safeguards are in place to provide qualitative and quantitative assurance of safety and reliability, societal and environmental wellbeing; and (last but not least) knowing who, how, why and when someone should be accountable.

If this is to be operationalised at a national scale, government departments and businesses need to be given permission (and a good nudge) to allocate resource, time, effort and budget to AI ethics, its risk management and impact assessment, its governance, and ultimately its compliance. It needs elapsing of time and experience to move from competence, capability, capacity building to maturity.

Ethics alone lacks "teeth" and it is the obligatory requirement and the enforceability that the law offers that make ethics translation into law so attractive. It moves voluntary codes of conduct and ethical principles to a stable and more sure footing when it is mandated and enforceable through legislation or regulation.

This needs a suitable national regulatory environment, that recognises how and where AI's impact interplays with existing law, and how legal and regulatory gaps can be plugged.

If this is to be operationalised at an international scale, AI ethics will need a common language and to be decolonialised. There is no one size fits all approach to this global ethical dilemma. This is a global–local problem and needs international cooperation and collaboration, but grassroots understanding of the problems it presents and the people it impacts in a given jurisdiction, sector or cultural space. The impact on the planet is a problem for us all, so making AI sustainable and handling the issue holistically so that we do not perpetuate existing environmental discrepancies and mismanagement through geo-political division.

International bodies and national governments being open, and regulators and regulated businesses being responsive will be key as we move from the age of AI discovery into the age of AI implementation.

The UN Human Rights Council's report [3], "The right to privacy in the digital age", attempts at identifying and clarifying principles, norms and best practices relating to the promotion and protection of privacy rights in the digital age that also addresses the responsibilities of businesses enterprises in this context. The report provides guidance on how to address the emerging and pressing challenges to the privacy rights in a pervasively digital world. It explores the trends and concerns that interfere with privacy from a growing digital footprint to state surveillance and describes the responsibilities of the states to recognise, respect and protect the citizens' rights to privacy and the necessity for oversight and safeguards. The report also defines responsibilities for the business enterprises including respect and observance of human rights and the underpinning policies and procedures appropriate to the context, size and nature of its operations including due diligence in identifying and addressing the impact of their operations on human rights. The UN High Commissioner for Human Rights makes a number of recommendations aimed at the states and the business enterprises in recognising, evaluating and addressing the full implications of new data driven/intensive technologies on the human rights of the citizens.

On the business front, there's a rather unexpected upturn trend in the environmentally sustainable and ethical index funds outperforming the traditional investment funds even considering the impact of the pandemic on the markets [4]. The ethical index funds are now regarded as mainstream, a position that has traditionally been regarded as niche and at best minority.

The ethical index funds launched by Vanguard [5], one of the biggest global fund managers are offered under the Environmental, Social, Governance (ESG) class of funds that are branded as aligned with investor ethical principles. These are the three categories of ethical criteria that sets ESG funds apart from the traditional high return regular index sectors and track specific stock market indices that exclude companies that do not meet the independently established ESG norms and standards. Similar trends are emerging in the climate focused funds.

Morningstar, the global research agency that examined 745 sustainable funds when compared with 4,150 traditional funds surprisingly found that the majority of ethical/sustainable funds matched or outperformed returns on traditional funds in the UK or abroad over multiple time horizons [4]. This situation continued even during the COVID crisis. Another interesting facet is the longevity of the sustainable funds, effectively doing better over longer periods without the quiet removal or merger with better performing funds that's practiced by fund managers to boost the overall performance figures.

Overall, funds with a better and robust environmental, social and governance (ESG) focus and strategic management, are seen by industry observers as responsible investment that are better performers financially thus having a positive bottom line impact whilst aligning with social and ethical values.

3. Need for a balanced approach

There is an increasing body of ex-ante AI risk and impact assessment lists and questionnaires and standards for internal quality control and best practice processes, and for ex-post AI audit.

In the teachings of the prophet Zoroaster (630–550 BC) the universe is portrayed as a battle ground for good and evil [6]. Taoism, that also emerged around the sixth century BC believe that ultimate reality is beyond the capacity of reasoning and rational thought and interpreted the changes in nature as a result of interplay between polar opposites of *yin* and *yang implying* a belief in the unity of opposites. In a similar analogy to the Zoroastrian forces of good and evil, the Taoists strive to attain and maintain a dynamic balance between the polar opposites of *yin* and *yang* which are seen as a spontaneous and innate tendency in all things.

The traditional approach to the identification, evaluation, and management of risks (potential losses arising from hazards) and rewards (gains and benefits arising from the exploitation of opportunities) is that of minimization and maximisation whereas these are essential attributes in any facet of life, as recognised and practiced by the ancient wisdom of Zoroastrianism and Taoism. A holistic and balanced approach to the understanding and rational impact assessment of Autonomous Decision Making and Algorithmic Learning Systems (ADM/ALS) is to treat the hazards and opportunities as intertwined and omnipresent albeit associated with inherent ontologic and epistemic uncertainties.

This holistic framework is shown in **Figure 1** where typically hazards and threats are transformed into a spectrum of potential risks and opportunities into rewards/ gains respectively [7]. The outcome is the spectrum and scale of risks and rewards that on balance informs the stakeholders in their desired and preferred decisions.

This framework provides a holistic, rational and unambiguous view of the key influencing factors in the impact assessment of ADM/ALS avoiding isolated treatment and confusing *upside* and *downside* terminology often employed to inadequately convey the same concepts or intent.

Such discussions of a holistic framework have led to an increasing need to ensure (and provide assurance of) oversight, to enable multi-disciplinary scrutiny of AI, to challenging the asymmetries of power between those organisations deploying AI systems (whether they be public sector or private sector), and those individuals, legal persons, people groups and wider society impacted by AI system outcomes. This has led the UK's ADA Lovelace institute to undertake a landscape review of algorithmic assessment and AI audit tools [8].

The challenge is that whilst seeking to balance tensions and trade offs and find tools, methods and approaches that can be operationalised to gain transparency

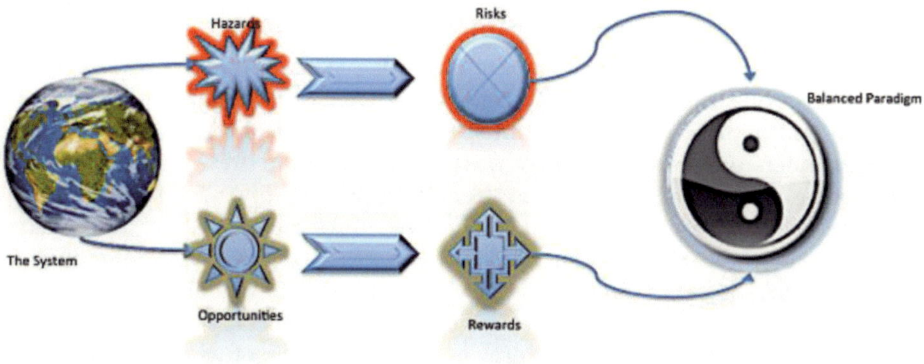

Figure 1.
A holistic risk–reward framework.

and hold AI systems (and the people behind them) to account, we need to find a common language, and an agreed taxonomy of terminology, that can not only cross language barriers but barriers of discipline too. There is still debate as regards how to define Artificial Intelligence itself. There is current debate as to whether ex-ante AI risk and impact assessments or ex-post AI audit should be looking at bias, unfairness or discrimination, or whether there is a case for all three areas to be within scope.

Ultimately the key is to ensure that no false positives or false negatives, no outliers or trends, no amount of data, no accuracy or inference, results in people being marginalised, excluded, rejected, or even expelled from operating and function in modern society. If people can or have been denied access to justice, social welfare, law enforcement, democratic engagement, employment, access to financial services, healthcare, education, or goods and services because of an AI system or as a symptom of all pervasive AI adoption without alternative or ability to opt out, we risk creating an ethical and societal divide. This requires algorithmic accountability, not least of all with the public sector [9].

In respect of the digital divide, according to the Office of National Statistics in the UK, in 2018 there were still some 5.3 million adults in the UK (amounting to 10.0% of the adult UK population) who were non-internet users. This simply provides a glimpse of the digital divide [10].

4. Pragmatic solutions

All tech solutionism aside, there is a place for human interventions, organisational approaches and socio-technical tools to develop and govern AI. There is no one size fits all approach. There is no one tool that can provide a silver bullet. It requires a holistic approach.

Understanding the purpose and outcomes to be achieved is a necessary first step. Many governments around the world are looking to algorithmic transparency to find ways of explaining automated decision making to its citizens. This on the one hand shows government to be open and accountable, but on the other hand it is a ruse to publicly legitimise their actions or inactions. Is not government responsible for the outcomes it creates in the public interest whilst also under a duty of care to ensure the safety of the wider public? If the public does not legitimise certain AI or ADM uses by government, what does that say to government in how it does or does not exercise its duty of care. How can we expect the government to fulfil its duty to the masses without leaving the less represented and marginalised groups in society exactly that....marginalised!

Transparency in all its forms is an important key step but must be accompanied with meaningful stakeholder engagement. Transparency is the gateway to many of the other ethical principles, but for transparency to do its work, it must be explainable and understood in context in a way which is relevant to the recipients of the information – the message received is after all the message given.

Tools such as AI registers and risk analytics platforms are needed to accompany governance but more need to be done. In order for there to be a holistic and pragmatic approach, AI governance need to take into account human intervention, organisational processes as well as technological tools, especially those that increase our understanding and provide meaning and interpretation of what exactly goes on in that opaque box. This way ethics can be turned into something that is operational. It also has opportunity to legitimise governmental use of AI and to reaffirm their societal mandate to act in the public interests.

5. Current trends and way forward

The European Commission has made a brave and bold move to seek to regulate in the area of AI. In an effort to build an ecosystem of excellence and trust, it seeks to preserve European values and protect the Fundamental rights of European citizens. It's human centred approach to AI is to be applauded, especially as it seeks to provide a governance structure for AI, with scope for risk and impact assessment, adherence to standards and other voluntary codes of conduct, providing for conformity assessment (akin to product liability legislation) for those AI deployments which are deemed "high risk".

Whilst this piece of legislation seeks to have extra-territorial effect like GDPR [X], it is not the GDPR of AI. Furthermore, it is a risk based, not principles-based piece of legislation like GDPR, but it does share something in common with GDPR: it is making the world's ears prick up. We may indeed see that all important "Brussels Effect" for AI governance crossing jurisdictional, geographical, and cultural divide, decolonising AI and AI ethics.

Barriers to global roll out and wider spread adoption of a regulatory approach such as this will be economic (determined by views of regulation stifling innovation), political (in the AI race), and will concern ethical disparities (public good versus equity and justice for the individual).

From a broader ethical perspective, three key areas of concern in development and deployment of ADM/ALS relate to Accountability, Transparency and freedom from unacceptable Algorithmic Bias. To this end, the IEEE-Standards Association has developed a suite of detailed criteria for evaluation, assessment and certification of these properties of ADM/ALS products and services under the "Ethics Certification Programme for Autonomous and Intelligent Systems" (ECPAIS). This programme [11] is a key facet of the IEEE-SA's Global Initiative and Ethically Aligned Design portfolio.

The three classes of ethical dysfunctions that may emerge in the embedding of ADM/ALS in products, systems and services require a systematic and credible independent evaluation and assurance to allay the public and private sectors' concerns and foster acceptance and deployment. To this end, IEEE-SA's suite of pragmatic and holistic certification criteria are now ready for deployment and tailoring for specific sectors and applications.

The high-level principles (Evaluation and Certification Factors) for each of the currently three ECPAIS suites are broadly defined as hierarchy of more detailed factors and criteria (typically, 10–20 for each of the depicted high-level factors) which are S.M.A.R.T i.e. specific, measurable, achievable, realistic and timely at the pertinent system or component level.

Transparency relates to the criteria and values embedded in a system design and the openness and disclosure of choices and decisions made for development and operation. This applies to the entire ADM/ALS context of application for the product or service under consideration such as data sets and not restricted to technical and algorithmic aspects alone.

Accountability considerations concern the commitment by individuals and institutions involved in the design, development or deployment of ADM/ALS to remain responsible for the behaviour of the system as long as its integrity is respected. This is predicated on the recognition that the system/service autonomy and learning capacities are the result of algorithms and computational processes designed by humans and those humans should remain responsible for their outcomes. A key driver in accountability is explicit, sufficient and proper documentation and traceability for system design, development and deployment.

Algorithmic Bias relates to systematic errors and repeatable undesirable behaviours in an ADM/ALS that create unfair outcomes, such as granting privileges to one group of users over others where they are expected to be neutral and unbiased. This can emerge due to many factors, from the design of the algorithm influenced by pre-existing cultural or institutional practices, the decisions relating to the way data is classified, collected, selected or used to train the algorithm, the unanticipated context of application and even presentational aspects emerging from search engines and social media.

The ECPAIS suites of ethics certification criteria are currently being extended to include ethical Privacy and tailored suites for high social impact domains including a bespoke suite for ethical assurance of COVID-19 pandemic related Contact Tracing Technologies [12]. This trend will continue to ensure ECPAIS embodies a broader and more comprehensive range of concerns in technology ethics.

6. Conclusions and the way forward

2021 should also be the age of AI ethics implementation, where operationalising AI ethics is not only seen as building trust to secure your customer base, an innovation enabler, providing legitimacy, and/or a competitive advantage, but an opportunity to build back better, recognising and addressing systemic inequalities and injustices, and creating a level playing field for people no matter who they are, their socio-economic circumstances, their background, or where they are in the world.

We ought not to consider AI and its application to the world in terms of an unlevel playing field but rather the world is the field in which everyone must play, how can we all work together best to create a playing field that everyone so that ALL can survive, be dignified and respected, thrive and flourish in using AI, with no one left behind [13].

In this endeavour, we ought to recognise that after two millennia of recorded civilization, consideration of ethics and social values in all that we do is a long overdue development. This therefore is a journey that thanks to the emergence of ADM/ALS we have just embarked on and should not be treated as a destination in line with many other facets and emergent properties of products, services and systems.

Author details

Ali Hessami[1*] and Patricia Shaw[2]

1 Vega Systems, United Kingdom

2 Beyond Reach – A Tech Ethics (Governance, Policy and Legal) Consultancy, United Kingdom

*Address all correspondence to: hessami@vegaglobalsystems.com

References

[1] UN Resolution. The Right to Privacy in a Digital Age, 26th September 2019, https://digitallibrary.un.org/record/3837297?ln=en

[2] The EU has sought after creating an ecosystem of excellence and of trustworthy AI: https://ec.europa.eu/info/strategy/priorities-2019-2024/europe-fit-digital-age/excellence-trust-artificial-intelligence_en

[3] Human Rights Council. The right to privacy in the digital age, Annual report of the United Nations High Commissioner for Human Rights and reports of the Office of the High Commissioner and the Secretary-General, A/HRC/39/29

[4] https://www.vanguardinvestor.co.uk/investing-explained/ESG-funds?cmpgn=PS0820UKBABES0001EN&s_kwcid=AL!11156!3!471139682346!e!!g!!vanguard%20esg%20funds&gclid=CjwKCAjwo4mIBhBsEiwAKgzXOLNIh1QevqqZ3sSHICDirehhFI5ZahQn_rOsfYJGHTW3xqeEnNaaXhoCUTsQAvD_BwE&gclsrc=aw.ds

[5] https://www.theguardian.com/money/2020/jun/13/ethical-investments-are-outperforming-traditional-funds

[6] Bekhradnia, S. (2007). The Beliefs of Zoroastraniasm, New Statesman, 8 January 2007

[7] Hessami, A. G. Perspectives on Risk, Assessment and Management Paradigms, Intech Open, ISBN 978-1-83880-134-2, 2019

[8] Examining the Black Box: Tools for assessing algorithmic systems, Ada Lovelace Institute, 29th April 2020, https://www.adalovelaceinstitute.org/report/examining-the-black-box-tools-for-assessing-algorithmic-systems/

[9] Algorithmic accountability for the public sector, Ada Lovelace Institute, dated 24th March 2021, https://www.adalovelaceinstitute.org/project/algorithmic-accountability-public-sector/

[10] Exploring the UK's digital divide – Office for National Statistics, dated 4th March 2019 https://www.ons.gov.uk/peoplepopulationandcommunity/householdcharacteristics/homeinternetandsocialmediausage/articles/exploringtheuksdigitaldivide/2019-03-04#the-scale-of-digital-exclusion-in-the-uk

[11] https://standards.ieee.org/industry-connections/ecpais.html

[12] IEEE Use Case—Criteria For Addressing Ethical Challenges In Transparency, Accountability, And Privacy Of Contact Tracing—Call For Global Consultation. https://beyondstandards.ieee.org/new-creative-commons-paper-addresses-ethical-hurdles-to-contact-tracing-adoption/

[13] Hutter, B. M. (2005). The Attractions of Risk-based Regulation: accounting for the emergence of risk ideas in regulation, ESRC Centre for Analysis of Risk and Regulation, Discussion Paper 33, London School of Economics & Political Science March 2005

Chapter 2

Technological Approach to Ensure Ethical Procurement Management

David Fourie and Cornel Malan

Abstract

Various studies have reported a positive connection between the public procurement and economic performance of a country, in terms of value for money, enhanced human welfare and improved economic growth. According to the World Bank, a distinction can be made between accountable governments where public procurement's share of the GDP is over fourteen percent, medium accountability countries with a share of thirteen percent and low accountability countries with less than twelve percent. In response to the ever-increasing complexity of procurement, many disruptive innovations as well as rapid developments in digitalization are reforming global supply chains. The principles of a sound procurement system include accountability, competitive supply, and consistency, which when viewed together with ethics and good governance, become the corners stones of an effective, efficient, transparent, and reliable procurement system. Ethical risks are possible in every stage of the procurement process; however, e-procurement has become a powerful tool to curb fraud, corruption, and unethical behaviour in public procurement as it reinforces the ethics of transparency, accountability, and integrity in procurement functions. With e-procurement being a relatively new form of procuring goods and services, it has been up against several challenges, notwithstanding the proven benefits of using electronic means in procurement. The movement to e-procurement has been a slow process globally, but various countries such as Germany, Korea, Brazil, and Zambia have already started to reap the fruits of their efforts. The main benefit of introducing e-procurement recorded by the World Bank has been a marked upturn in transparency and competition. This chapter aims to unpack the link between technology, procurement, and ethics towards the provision of goods and services by governments for the greater good of all.

Keywords: public procurement, public service delivery, information technology, ethics, e-procurement

1. Introduction

Public expenditure and- procurement form a major part of a country's economy and is considered an important indicator of the efficacy of a government, given the direct link of such to public service delivery - policy instruments that governments use. Studies showed a connection between public procurement and economic performance that is reflected in a country's economy and citizen's well-being [1]. Public procurement is a factor when measuring productivity of the public sector [2]. Wagner's law argues that economic growth leads to an increase in public expenditure. Thus, there is a correlation between public spending and a country's economic development [3]. The essence of public procurement is to achieve value for money, which manifests in enhanced human welfare and improved economic growth [4]. Sound practises is important for countries' international trade and foreign investment. Public purchase of goods and service is estimated to be 13.3% of the Gross Domestic Product (GDP) of European countries [5]. The GDP also reflects government's ability to deliver services. In fragile states less than 5% is spent on public procurement. The World Bank makes a distinction between accountable governments where public procurement's share of the GDP is over 14%, medium accountability countries with a share of 13% and low accountability countries with less than 12% [6].

The International Trade Centre submits that "governments are market regulators and market participants" and as such legal and regulatory frameworks are established while spending trillions of dollars to purchase goods works and service to fulfil their public functions [7]. Public procurement is a key economic activity that is used by governments for amongst others the attainment of horizontal or secondary objectives such as supporting Small Macro Enterprises, protection, and advancement of previously disadvantaged social groups (which includes empowering of women), and to stimulate innovation as well as green technologies. There are mainly two opinions from opponents of secondary objectives. Firstly, public institutions should only aim to achieve value for money and timely delivery of goods and services in these times of tight budgets. Secondly, the creation of unnecessarily complex procurement process will increase the cost of the procedure, compromise the primary objective disproportionately and reduce competition [8]. Government's accountability responsibility is not only towards the public whose money is spent but also disappointed tenderers and potential suppliers. For this reason, procedures and practices must be developed that can stand up to scrutiny. Public institutions achievement of effective and efficient procurement objectives and results are dependent on three elements. Firstly, setting procurement standards, specifications, objectives, and goals and achieving them. Secondly, ensure satisfaction of all role players and lastly, enforcing applicable procurement policies and regulations [9].

In response to the ever-increasing complexity of procurement, many disruptive innovations as well as rapid developments in digitalization are reforming global supply chains. The current modus operandi of the procurement function within and between countries, in both the private and public procurement environments is being challenged to adapt accordingly, to be able to align procurement to aspects such as transactional automation, proactivity of supplier relationship management and predictability [10]. In addition, e-procurement has become a powerful tool to curb fraud, corruption, and unethical behaviour in public procurement as it reinforces the ethics of transparency, accountability, and integrity in procurement functions [11].

2. Procurement contextualised

The primary function of public procurement is "procuring goods and providing services and infrastructure on the best possible terms" [11]. The definition can be broadened by distinguishing between procurement and public procurement. "Procurement means obtaining goods, works, consultancy or other services through purchasing, hiring or obtaining by any other contractual means; and public procurement means procurement by a public body using public funds" [12]. "Public procurement processes are sequences of activities starting with the assessment of needs through awards to contract management and final payment" [13].

In most countries public procurement follows a cycle based on legislative requirements, administrative processes, and budget timelines. Legislation requires fair and transparent process where competition adds to decreased cost and increased quality [14]. Policy developers are challenged by dilemmas such as: develop flexible structures yet maintain accountability and control, limit opportunity for fraud/mismanagement while reducing operational constraints, increase economic efficiency while satisfying political demands, increase open and transparent competition while achieving best value, and applying best practices while confronting legal limitations [15].

2.1 Principles of procurement

In general, principles of a sound procurement system include accountability, competitive supply, and consistency, which when viewed together with ethics and good governance, become the corners stones of an effective, efficient, transparent and reliable procurement system [11].

In order to avoid a lack of integrity, an all-inclusive approach to risk mitigation and prevention of corruption through the entire public procurement process is essential. Integrity refers to the protection of ethical norms and standards relating to the principles of "honesty, professionalism and righteousness" [16], and as such provides the foundation for guaranteeing that public procurement processes are fair, compliant and non-discriminatory in nature and application.

By only addressing integrity issues in one step of the process may result in risks in some other stages or mitigating only one type of risk may give pave the way for integrity infringements through other methods. For example, measures aimed at ensuring compliance during the initial bidding phase when the procurement needs are being determined, may not completely prevent political influencing. In a similar fashion, declarations of interest by procurement officials may not necessarily prevent large scale so-called "bid rigging" or small fraudulent actions from occurring [16]. In adopting such a holistic approach, the Organisation for Economic Co-operation and Development (OECD) Recommendation on Public Procurement places emphasis on mutually supportive principles which may, "directly or indirectly, prevent corruption and stimulate good governance and accountability in public procurement" [16] and include:

- Integrity
- Transparency
- Stakeholder participation
- Accessibility
- E-procurement
- Oversight and control [16].

2.2 The stages of the procurement process

The United Nations published a flowchart reflecting various stages in the procurement process [17], as illustrated in **Figure 1** below. Directives or standard operating procedures provides guidance to procurement officials that acts as a reference tool for coordinated and integrated actions that inherently leads to the

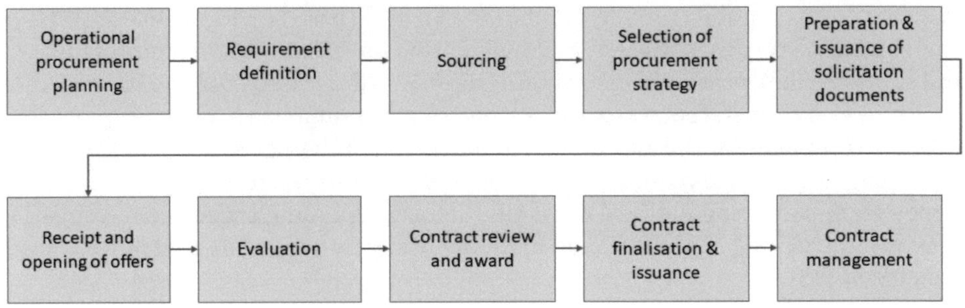

Figure 1.
The stages of the procurement process [17].

attainment of an open, efficient, effective, and transparent process. Directives are unique to procuring entities that largely depend on their vision, mission, and service delivery mandate.

The acquisition process commences when a societal need is identified, and availability of funds is confirmed. Operational procurement planning clarifies what is needed by who at what time as well as the quantity and quality of goods and services required. Failure to plan effectively may result in service delivery goals not being achieved [17]. Operational procurement, requirement definition, sourcing, selection of procurement strategic and preparation and issuance of solicitation documents are elements of Demand Management where consolidated procurement plans are annually developed for the whole institution. During demand management government entities must take into consideration social, economic, and environmental aspects that may have an impact on appropriate and sustainable service delivery. The demand plan is aligned to the institution's approved strategic plans and procurement may not take place outside of the approved plan, except in cases of emergency. To allow for sound procurement principles and prudent financial spending institutions would have standard operating procedures in place for emergency procurement.

Cost effective procurement are dependent on accurately estimated costs which will at a later stage determine the procurement method. During the planning stage government institutions need to consider secondary policy objectives that will impact on drafting of specifications and the selection of an appropriate procurement strategy. Secondary policy objectives are defined as objectives focussing on promoting innovation and new ventures such as small and medium-sized enterprises, or objectives aiming at increasing sustainability. Secondary objectives could also be to illustrate social accountability or to support other wider policy objectives aiming for economic growth, given the increasing use of procurement by many governments to promote social welfare, in addition to the primary procurement objective of providing goods or services [18]. Fairness, value for money [competitiveness and cost-effectiveness] and transparency are internationally regarded as the cornerstones of good procurement practices [19]. For this reason, most countries utilise a committee system. The Bid Committee must be constituted with knowledgeable members where specifications are not treated as hindrances to access the market but as defining what the government wants to buy. In order to abide to the principles of fairness and competitiveness, and to avoid the possible chance of favouritism or exclusion of entities in the bidding process, procurement entities are prohibited from including any specific reference to a certain process, or inclusion of descriptions of products or goods by using a specific model, type or make, the registered product name or brand, production location of methods used, in their bid specifications [20].

Although their names may vary, there are generally six methods on how the market can be approached. The first being open tendering or competitive bidding that encourages effective competition but is not ideal for large or complex works. Open tendering lends itself for maximum competition where the potential of favouring a specific tenderer is lower and collusion is less likely. The disadvantage being overall cost and time-consuming. Restricted tendering also known as 'selective tendering' is a method where limited agencies are invited by the procuring entity. The process is still considered to be fair and transparent when the entity pre-determined a set of guidelines for invitation. The potential for corruption is higher due to exercise of discretion. Another method of procurement is Request for Proposals (RFP) where prospective suppliers are requested to propose why their goods or services are the best for a specific institution's need. Two Stage Tendering takes place when suppliers are requested to provide two envelopes. The one envelope containing the proposal and one with the financial information. Selection takes place in two phases. First the best proposal is selected and then matched to ascertain whether it is also the best price. Request for Quotations are used for small-valued goods or services. The advantage being that it is fast with limited paperwork. The last method, Single-Source selection is non-competitive and requires prior management approval. This method is used when the procuring entity pre-selected a sole provider due to emergencies, if only one supplier is qualified, where the product or service are only available from one supplier or when it is a continuation of existing work [21]. Issuing of solicitation documents serves as an invitation for prospective or interested bidders to participate in the process. Fair access to solicitation documentation was restricted until most countries implemented e-procurement. The advantages and disadvantages of e-procurement will be discussed later in the chapter.

The next major stage in the bidding process is where offers are received and opened, evaluated in terms of compliance with minimum qualification criteria and awarded. Ethics and fair dealing are of critical importance during this stage in order to maintain public trust. Tender Evaluation committees must be constituted of members who has been cleared of a potential conflict of- interest (e.g., through mandatory disclosure or declaration) [21] and are competent and knowledgeable officials in accordance with the general law of equal treatment, non-discrimination, transparency, and confidentiality of information [22]. The following activities will usually be carried out; formal compliance check (submission of solicitation documents), technical and substantive compliance check (evaluation against advertised specifications), choice of the best tender on the basis of the advertised award criteria and recommendation for the award of the contract. The main objective of bid evaluation is to ensure that the lowest acceptable bid has the necessary qualification, experience, and staffing to perform the contract [23]. However, the choice of a successful tenderer should be based on value for money such as the most optimum tender and not only the lowest bidder. The aforementioned contradicts the misguided belief that price is the determined factor. Recommendation for award is made to a third committee, the Bid Adjudication committee.

Bid adjudication committees in most instances comprise of senior officials and as is the case with the other committees, the members are appointed in writing. For the sake of unbiasedness and transparency, members of the evaluation committee should not be a member of the adjudication committee as well. Bid specification committees compile specifications, bid evaluation committees evaluate responses against the published specifications and the bid adjudication committee approve that a tender may be awarded to the recommended bidder. The adjudication committee is not mandated to select a tender but to confirm that the process followed was fair, transparent and that the product will satisfy society's needs and contract commencement can follow. Clear separation and segregation of functions should allow for integrity in the public procurement system. All matters prior to award is of an administrative nature and post tender award legal rectification is sought when so required [22].

2.3 Using functionality during procurement

The use of functionality in procurement refers to the application of predetermined evaluation criteria, as per the tender specifications, aimed at evaluating the operational, technical, and practical ability of a tenderer to provide goods or services as required in terms of specific aspects such as the quality of the goods, or the dependability, sustainability, and durability of a service [24]. The evaluation criteria for measuring functionality must be objective and suppliers must be informed of the following: (i) the evaluation criteria for measuring functionality; (ii) the weight of each criterion; (iii) the applicable values; and (iv) the minimum qualifying score for functionality [25].

Certain aspects must also be taken into account, in the event that an institution invites a bid which will be evaluated by means of specific functionality criteria. These include:

- the specific functionality evaluation criteria;
- the specific weight of each criterion during evaluation of the bid;
- the relevant value which will be allocated during the bid in terms of the performance of the tender against the criteria; and
- the minimum score required in terms of the functionality criteria [25]

Functionality criteria requires bidders to obtain a predetermined minimum score for the functionality criteria, in order to be considered for further appraisal during the award stage, with regards to price and preference. Functionality evaluation criteria must be stipulated as such in the bid documents and may require additional information such as proof of previous performance or applicable experience, credentials of the envisaged personnel to be involved, or how the tender intends to ensure the necessary skills transfer during the project, etc. The specific weighting allocated to each criterion must be bid-specific – a generic or blanket approach does not justify the use of functionality. in the same manner, the scoring for each criterion should also be specific to the tender requirements to be viewed as objective [24].

3. Ethics and governance in procurement

3.1 Contextualisation of ethics

Ethics is defined by the Oxford dictionary as "moral principles that govern a person's behaviour or the conducting of an activity" [26]. In a public sector context, ethics is defined as "standards that guide the behaviour and actions of public officials in public institutions and (that) may be referred to as moral laws" [27] while ethical behaviour is described as behaviour that is "not only good for oneself, but also good for another" [28]. Hence, it can be argued that, within the context of a public sector supply chain, ethical behaviour necessitates a person not only to act in the best interests of the particular institution involved, but also to adhere to certain stewardship criteria [29].

Public procurement is an important tool towards the well-functioning of government and delivery of services which in turn are fundamental for development, growth, and improved social welfare in any country. Sadly, public procurement is one of the government activities most exposed to acts of corruption, fraud, and mismanagement, given the substantial amounts of public funds used for public procurement [16, 30]. Over and above the sheer volumes of transactions and the financial interests involved, corruption risks are aggravated by the intricacy of the procurement process, the array of stakeholders and the close interaction between public officials and external entities [16].

Various types of corrupt acts may take advantage of these weaknesses, such as "embezzlement, undue influence in the needs-assessment, bribery of public officials involved in the award process, or fraud in bid evaluations, invoices or contract obligations" [16]. Ethical risks are therefore possible in every stage of the procurement process, and red flags include undue influence, conflict of interest, and various kinds of fraud risks, as illustrated in **Figure 2** below.

Non-compliance with Supply Chain Management (SCM) policy and regulations is identified by Fourie and Malan as challenges in Public Procurement [11], with

Pre-tendering phase	Needs assessment and market analysis	• Lack of adequate needs assessment • Influence of external actors on officials' decisions • Informal agreement on contract
	Planning and budgeting	• Poor procurement planning • Procurement not aligned with overall investment decision-making process • Failure to budget realistically or deficiency in the budget
	Development of specifications/ requirements	• Technical specifications are tailored for a specific company • Selection criteria is not objectively defined and not established in advance • Requesting unnecessary samples of goods and services • Buying information on the project specifications.
	Choice of procurement procedure	• Lack of proper justification for the use of non-competitive procedures • Abuse of non-competitive procedures based on legal exceptions: contract splitting abuse of extreme urgency, non-supported modifications
Tendering phase	Request for proposal/bid	• Absence of public notice for the invitation to bid • Evaluation and award criteria are not announced • Procurement information not disclosed and not made public
	Bid submission	• Lack of competition or cases of collusive bidding (cover bidding, bid suppression, bid rotation, market allocation)
	Bid evaluation	Conflict of interest and corruption in the evaluation process through: • Familiarity with bidders over time • Personal interests such as gifts or future/additional employment • No effective implementation of the "four eyes-principle"
	Contract award	• Vendors fail to disclose accurate cost or pricing data in their price proposals, resulting in an increased contract price (i.e., invoice mark-ups, scope creep) • Conflict of interest and corruption in the approval process (i.e., no effective separation of financial, contractual and project authorities/ lack of disclosure) • Lack of access to records on the procedure
Post-award phase	Contract management/ performance	Abuses of the supplier in performing the contract, in relation to its quality, price and timing: • Substantial change in contract conditions to allow more time and/or higher prices for the bidder • Product substitution or sub-standard work or service not meeting contract specifications • Theft of new assets before delivery to end-user or before being recorded • Deficient supervision from public officials and/or collusion between contractors and supervising officials • Subcontractors and partners chosen in an on-transparent way or not kept accountable
	Order and payment	Deficient separation of financial duties and/or lack of supervision of public officials leading to: • False accounting and cost misallocation or cost migration between contracts • Late payments of invoices • False or duplicate invoicing for good and services supplied or even worse, not supplied, and for interim payment in advance entitlement

Figure 2.
Possible risks per procurement phase [16].

one of the challenges being unethical conduct of procurement officials. They view unethical conduct in the same light as acts of corruption, fraud, nepotism, and bribery, all of which have an adverse bearing on the service delivery quality levels. Munzhedzi investigated whether procurement processes and- practices and corruption are inseparable twins. He is of the opinion that whenever one of the two is mentioned, the other one has to follow in the next line. The illusion that rules have been respected is important to the transgressor in order to reduce the risk of being caught [31]. Instances are found where the composition of bid adjudication members is manipulated to serve the purpose of those who do not wish to act with integrity. For example, a senior supply chain practitioner is replaced with a junior official that can easily be intimidated to agree with the decisions of the rest of the committee. At every stage during the procurement process, procurement is a prime candidate for corrupt activities, cronyism, favouritism as well as bribery. During bid evaluation a more expensive offer can be selected based on 'legitimate reasons' however corrupt bid rigging took place behind the scenes [18] that in most instances will not stand up in a court of law. In most instances it is not the procurement system that failed the most vulnerable in society but rather greed of those in positions of trust. Patras [32] agrees that "errors", or infringements of the rules, can also happen. Human errors for example incorrect calculations of assessment criteria, take place with public funds and has the same financial impact as corrupt practises [32]. Procurement should be viewed as a profession where governments ensure that procurement officials meet high professional standards of knowledge, skills and integrity [21].

3.2 Combatting corrupt activities in public procurement

The fight against public procurement corruption is a global phenomenon. Governments have become reliant on prescriptive procurement standards, exclusions, and formal processes that according to the World Economic Forum increased the costs of corruption instead of preventing it from happening. "Limited political will" is identified as the main culprit for the lack of efficacy of preventive actions. The World Economic Forum advocates for "technologically induced sunlight" in an effort to disinfect procurement processes [33]. One of the policy objectives of a procurement approach should be to deter, detect, and punish corruption, versus the objectives of public procurement which are fairness, equity, transparency, competition and achieving value for money. These objectives though complementary can represent a trade-off or "dual challenge" for good governance as integrity in the form of fairness and transparency, is crucial for a successful public procurement system whilst also aiding in the uncovering of corrupt activities [30].

Integrity standards being a core element of professionalism, influences the behaviour of procurement officials and contributes towards the creation of a culture of integrity [21]. It is submitted that procurement officials have a responsibility to familiarise themselves on legislation, regulations, and standard operating procedures. Ignorance towards procurement regulations by procurement officials is no excuse as no one is deemed ignorant of the law. McCue and others have a different view of the impact of a proverbial "stick". The stick being rules and processes that limit discretion. They are of the opinion that procurement professionals might struggle with what they perceive as the right choices and what is defined by other as the right choices [34].

The OECD maintains that protection of integrity should be the point of departure to prevent corrupt activities in public procurement. Public procurement policies as well as role-specific standards and codes of conduct, for procurement officials are essential. It is especially important to prevent officials' private interests from having any inappropriate or illegal sway on their performance of their public duties and responsibilities and strict regulations prohibiting such must be in place. In fact, research indicates that the majority of conflicts of interest are linked to either an opportunity to gain in terms of personal, family or business interest, promises of possible career opportunities, donations or presents, and the failure to disclose confidential information relating to such activities [29].

Regulatory compliance to public procurement is largely dependent on ethical behaviour where ethics is a pre-requisite for reducing non-compliance. Sarawa and Mas'ud [35] developed a mediation model that identifies four determinants that impacts on public procurement regulatory compliance. The first determinant being 'professionalism' is closely related to cognitive development of an individual whereby training and professional development will manifest itself through moral obligation [ethical behaviour] which will ultimately direct an individual to comply with public procurement regulations. The second determinant for compliance is 'familiarity' which can be achieved through on the job learning and reading of procurement regulations. According to Sarawa and Mas'ud [35] the third determinant, enforcement' "will make public procurement officers behave ethically and consequently comply with public procurement regulations." The fourth and last determinant is 'resistance to political pressure'. They are of the opinion that public procurement officers should evaluate the legitimacy of instructions from powerful forces and resist any pressure towards violation. Persistent ethical behaviour will eventually ensure compliance with prescribed regulations [35].

A SCM Code of conduct has been implemented in a number of OECD countries which focuses in detail on preventing conflicts of interest. In addition, training programmes are also in place to increase the levels of ethical behaviour and public accountability. Furthermore, the OECD advocates for the promotion of transparency, by means of provision of public access to information, and timely and effective responses to information request, as means of protecting public interest and procurement integrity. It is presumptuous to assume that the relationship between maintaining transparency and increased integrity is guaranteed. It is in fact reliant on a number of requisite and enabling factors to be in place for accountability to be truly effective, and includes public access to data, accurate, reliable, and timely feedback, quality data, regular and comprehensive reporting as well as effective whistle-blowing processes and methods, including the protection of whistle-blowers [16, 29].

External stakeholders' participation during the procurement stages has also proven to lead to increased transparency and honesty, given the increase in public scrutiny, provided that privacy or confidentiality, equality in treatment, as well as other legal requirements in the procurement process are adhered to [29].

4. The use of technology in procurement

The utilisation of information and communication technologies (ICT) in public procurement can promote increased transparency and facilitate ease of access to public tenders, in fact some argue that technology is essential to the modern-day public procurement in terms of fair competition, equity and transparency [36]. E-procurement, which is the application of information and communication technologies in public procurement, increases public procurement transparency, enables greater access to public tenders, increasing outreach and competition, decreases direct contact between procurement officials and enterprises, and provides for easier uncovering of irregularities and corrupt activities, such as bid manipulation [16]. Digitalisation reinforces internal corruption prevention controls, strengthens the early detection of integrity infringements, and provides audit trails for investigation purposes [29].

Fraud, corruption and unethical behaviour are serious concerns badgering public institutions. E-procurement is globally considered to be a reorganisation or procurement reform process of goods and services procurement in an effort to prevent or reduce corruption levels and improve efficiency. Transparency International uses the following definition to explain information and communication technologies: "the use of any internet-based inter-organisational information system that automates and integrates any parts of procurement process in order to improve efficiency, transparency and accountability in the wider public sector" [37]. Added advantages are reduced cost, increased accessibility of information and automation practices prone to corruption. Automation of basic steps in the bidding process such as distribution of forms and acceptance of documents eliminates direct human interaction there by limiting opportunities of manipulation and requests for bribes and kickbacks [38].

Transparency and openness in public procurement contributes to more efficient allocation of resources, investment is attracted by lowering risk, efficiency of local suppliers is increased due to fair competition, and local suppliers may become competitive exporters. Advances in technology can be used as a key to respond to transparency demands from both the public and prospective suppliers. Agwot in [39] considers e-procurement to be a double sword principle; on the one side it contributes to the principles of transparency, value for money and fairness while increasing private practitioners' confidence in the procurement system [39].

The public procurement process is at its most vulnerable during tendering and contract award. Influential contractors use coercive power to get a contract. Other contractors are not able to submit tender documents due to coercive threatening and/or government officials who are indirectly involved for their own benefit. Online bidding through e-procurement reduces cartels and bid rigging amongst bidders. Poor, infrequent auditing that lack co-operation with other relevant agencies are deterrents for transparent and effective flow of information. E-procurement centralises data to improve audit and analysis [38]. Not enough attention is given to contract implementation. Corrupt practices take place through fraudulent invoicing, overbilling, under performance, and failure to adhere to specification standards. The use of e-procurement facilitates easier control and oversight over the procurement cycle through standardised and streamlined processes.

4.1 Link between technology, procurement, and ethics towards the provision of goods and services by governments for the greater good of all

A series of major technological shifts has transformed the global procurement environment in recent times, given the need for lower computing costs, higher volumes of data mining and storage, superior forecast precision, dependence on reliability of data, and evaluation of supplier performance. As the needs of governments become more intricate, technological development have aimed to streamline the procurement function, towards higher levels of consistency, automation of manual transactions and improving the management of the supplier relationship [40].

Governments acquire goods and services to support their operations in providing public services. Services include amongst others, security, transportation, educational systems, medical services, and infrastructure. Service delivery is hampered by unethical conduct where only a few benefits from government's service delivery responsibilities. Unethical business practices are heightened in the absence of transparency. Increased performance and service delivery can be achieved where systems are introduced that allows for transparency, fair treatment in bidding and awards of tenders, accountability and responsible purchasing and supply [41].

Electronic procurement has been in use in various countries for a number of years, in different ways, each with its distinct advantages [42]. The OECD and its member States developed a "Compendium of good practices for integrity in public procurement" [43]. The Compendium recognises opportunities for the use of technology during the various procurement stages for the greater good. The online publication of relevant public procurement information results in increased confidence in procurement and leads to more competition. One example is found where the Argentinian government not only publishes procurement opportunities, the timelines for submitting, selection and evaluation criteria, but they also provide a platform for "common questions" regarding the functioning of the public procurement system. Another example is the use of a central procurement system by the government of Mexico, thereby creating empowerment mechanisms for society, their civil organisations, and the watchdog media to examine government procurements at depth and in detail [43].

Corruption during the project planning phase takes place when an unwanted project for private gain is planned by a senior official or political office bearer. Spending public money on unwanted projects does not fall in the ambit of good governance. Such projects are likely not appropriate to society's needs and therefore seldom sustainable. Confidential information may be leaked, or unnecessary qualification criteria is added. E-procurement plays an anti-corruption role by providing the public or potential bidders with a platform to view and monitor procurement activities through the government web portal. Replacement of paper-based procedures with digital technology-based communication does however not prevent unethical decisions outside of the approved demand plan. During product design, documentation specifications can be compiled to favour a particular supplier, or the prevalence of corruption is concealed in an unnecessary complicated tender. Transparency as a corruption antidote is applicable when project specifications can be viewed on a web portal [38].

The primary procedure most frequently used for the acquisition of goods and services by government, is Competitive Bidding, given its competitive nature and value for money results. The electronic version of competitive bidding is known as the so-called electronic reverse auction procedure [42]. The United Nations Commission for International Trade Law's (UNCITRAL) Model Law on Public Procurement of Goods, Construction and Services Article 2 describes electronic reverse auctions as an on-line purchasing method which, is used by the procuring entity in a present or real-time state to decide on successful tender proposals, and entails suppliers or contractors presenting lowered bids during a scheduled period of time and the automatic evaluation of bids until a successful bidder is selected. It therefore requires active bidders' participation during both bidding and award processes [44]. A traditional means of bidding, the English Auction, occurs when bidders gather at a determined location and publicly reveal their bids to the other bidders. As reserve prices are continuously increased, a bidder is permitted to bid several times until no other bidders increase and only one bidder remains claim the item at the last bid price [44]. The same principles apply during electronic reverse auctions; however, the bid values are disclosed electronically, and the values of bids are decreased instead of increased. Points for various aspects of the bids are awarded based on the results of a mathematical formula. The UNCITRAL Model Law requires the procuring entity to publish an invitation to tender electronically, which must stipulate the subject matter of the procurement, the contractual terms and conditions, examination criteria of the bids, including the mathematical formula to be used, as well as the evaluation procedure. The invitation must further inform bidders whether any other component than price will be evaluated, such as preference or quality [44].

Governance in the form of procurement oversight and control is crucial to ensure accountability and is promoting integrity in the public procurement process. In addition, valuable evidence on the performance and efficiency of the procurement cycle is to be gained from these processes. The governance process should be based on an environmental risk analysis of the government process to ensure an adequate oversight and control system, which will be supported by observations from oversight and control activities, in terms of new and/or emerging risks or red flags, thus allowing for continuous updating and adjustment of the oversight and controlling system [10].

Likewise, suitable sanctions in the form of consequent management actions for illicit behaviour revealed by management control actions, may serve as a useful warning to those contemplating corrupt actions. Management control and oversight are fundamental pillars of the OECD instruments underpinning the utilisation of comprehensive governance systems in the public sector, more specifically in the public procurement domain. The OECD Recommendation of the Council on Public Integrity promotes a so-called framework for control and risk management to preserve public sector integrity, through:

- enabling a procurement environment controlled by clear and fair objectives, indicative of senior officials 'commitment to upholding public service values and principles of integrity and, which provides for a sufficient assurance of an organisation's performance, its effectiveness, and its compliance;

- guaranteeing a strategic risk management approach; and

- ensuring consistent control mechanisms, inclusive of clearly defined and applicable procedures to provide for reporting of plausible suspicions of violations of laws and regulations, without fear of reprisals, as well as the facilitation and investigation thereof [10].

The use of blockchain technology in procurement has also been found to be effective in managing integrity of the process [45]. A blockchain is described as "an open, decentralized, and distributed digital ledger that is used to create a trusted peer-to-peer network for exchanging information, value, and assets across many computers (i.e., nodes)". A blockchain records and encrypts time-stamped transactions between the involved partners which are then unassailable or unchangeable. These transactions are grouped in a bundle or "block" in a chronological and logical and order [10]. The use of blockchain thus creates new opportunities to drive increased procedural integrity and authenticity through the use of a technology driven decentralised platform for validating data, information, and transactions, independently of any third-party control in a transparent, verifiable, secured and lasting format. One of the important features of blockchain in procurement is that the different SCM and procurement stakeholders involved in the procurement processes have access to an unalterable, secure and shared database in that under one platform, entities can access the accumulated data of suppliers' portfolios, various services level agreements, project details etc. Refer to **Figure 3** below for leveraging aspects to be obtained from using blockchain technology [10].

4.2 Examples of successful E-procurement practices

Effective internal controls are intended to guarantee efficient public procurement processes while at the same time achieving integrity-related goals and objectives. Internal controls in procurement are intended to verify the degree of compliance to legal, financial, and administrative procedures, and include internal audit activities as well as management and financial controls. Furthermore, coherent internal control practices across the public sector safeguard the consistent application of procurement rules, regulations, and standards [16].

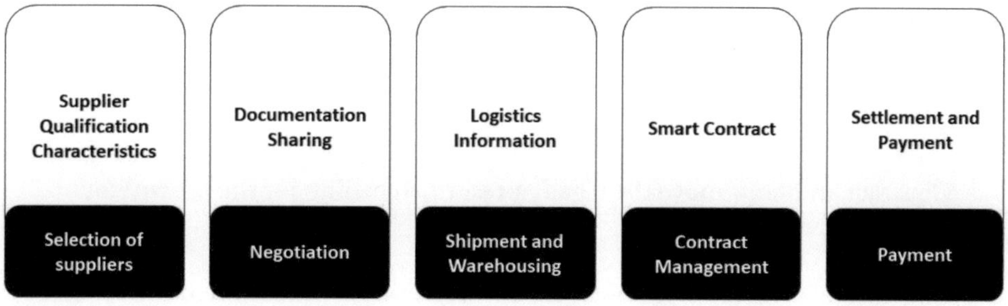

Figure 3.
Leveraging procurement process with blockchain technology [40].

4.2.1 Electronic workflow: processing and tracking information on public procurement in Germany

The Federal Procurement Agency in the Ministry of the Interior in Germany (FPA) has implemented an electronic workflow process which aims to centralise all procurement related information of the FPA and record the outcomes of ongoing procurement procedures during the different stages. The FPA maintains a document management system to ensure transparency as well as provide for an audit trail of all procurement decisions, investigations of suspicious transactions and also allows for the department to apply quality management to examine documents randomly or systematically in the system, while the internal audit teams can review previous transactions that have been identified as having a higher corruption risk. These inspections are not only to prevent or detect corruption, but also to ensure economically beneficial public procurement practices [16].

4.2.2 Public spending observatory in Brazil

E-procurement plays a key role in public procurement practices in Brazil, which is reliant on advanced technology, both internally, in terms of management and-database systems, as well as externally, for example, the Internet. E-procurement is of strategic importance at the Federal level for producing efficient and transparent results and maintaining effective control over public expenditure [36]. In 2008 the Office of the Comptroller General of the Union launched the Public Spending Observatory (Observatório da Despesa Pública) towards continuous detection of misconduct and corruption and appropriating the relevant sanctions. Through the Public Spending Observatory, procurement expenditure data is being cross-checked or verified against other government databases, to identify possible situations that, while not per se confirming any irregularities, may justify further investigation. Using data analytics and historical trends, daily cross-checking actions are performed between the procurement database and other government data, and this cross-checking exercise produces so-called red or orange" o flags, which are followed up and investigated by officials within the Office of the Comptroller General of the Union, if so required. Follow-up activities are often conducted in conjunction with special advisors on internal control as well as internal audit units to pinpoint possible procurement irregularities, such as:

- Business relations between suppliers participating in the same procurement procedure.
- Personal relations between suppliers and public officials in procurement procedures

- Non-compliance by suppliers with tender submission deadlines.
- Supplier's bid submissions or company records with the same registered address.
- Contract amounts above the legally prescribed ceiling for the procurement modality used.
- Contract amendments above an established limit, in violation of the specific tender modality.
- Micro- and small enterprises linked to other enterprises.
- Micro- and small enterprises with shareholders in other micro- and small enterprises [16].

The Transparency portal hosted by Brazil's government offers real-time access to people on information of budget execution and direct monitoring of government programs [43].

4.2.3 Integrated E-procurement system KONEPS in Korea

The digitalisation of procurement processes reinforces internal anti-corruption controls and detection of integrity breaches, and it provides audit services trails that may facilitate investigation activities. In Korea, the implementation of a national e-procurement system has resulted in a significant improvement in the transparency and integrity of the public procurement. In 2002, the central procurement agency of Korea, the Public Procurement Service (PPS), introduced a fully integrated, "end-to-end e-procurement system" called KONEPS, which provides for all aspects of the procurement cycle such as one-time registration, tendering, contracts, inspections, payments, as well as electronic exchange of documents. KONEPS links a multitude of external systems to share and retrieve information, and runs a one-stop service, including "automatic collection of bidder's qualification data, delivery report, e-invoicing and e-payment" [16]. According to PPS, the system has boosted efficiency, considerably reduced transaction costs, resulted in increased participation in public tenders and has considerably improved transparency. In addition, the Korean Fair-Trade Commission makes use of an automated KONEPS system for detecting suspicious bid strategies (named BRIAS), which has sharply decreased instances of corruption by preventing illegal practices and collusive acts. So-called 'Cover quoting', during which a tenderer submits more than one tender is illegal and to prevent this from happening, the government of Korea introduced "Fingerprint Recognition e-Bidding", limiting a tender to only one per company. The government also eased market accessibility through the use of smartphones during the bidding process [16].

4.2.4 Zambia's public E-procurement

E-procurement has not been widely used in by governments in Africa to date. According to the World Bank the three main reasons for the slow adoption of e-procurement in Africa have firstly been a low level of capacity building by African governments, secondly a lack of ICT infrastructure as well as limited internet access bandwidth challenges; and lastly, archaic administrative processes and restrictive cultures in African governments [44].

Policy makers in the Zambian government published the Public Procurement Regulations in 2011, under the strict provisions of the 2008 Public Procurement Act, under the auspices of the Zambia Public Procurement Authority (ZPPA), an autonomous regulator mandated to regulate procurement of services, goods and works in Zambia. Since the publication, research indicates that public entities and local authorities have been in support of a move to e-procurement as it is expected to yield several advantages, such as reduced opportunities for corrupt practices, access for prospective bidders to tender opportunities whether residing within the country or abroad, and compliance to public sector procurement legislation, thereby preventing service providers or procuring entities from deviating from the procedures set out by ZPPA [46]. E-procurement is also less expensive in terms of "hardcopy bid filing" which was required for the previous centralised traditional filing system, as well as saving money for participants in terms of lodgings in and transport to Lusaka. However, there are some concerns regarding non-availability of internet facilities or as a result of connectivity issues preventing possible participants in deep rural areas from submitting their electronic bids [47].

5. Advantages of E-procurement

It has been noted that implementing e-procurement is potentially valuable in guaranteeing a more efficient public procurement process which results in reduced costs for goods or services, a decrease in procurement cycle time periods, administrative cost and improving communication. The active involvement of bidders during a real-time procurement, being continuously informed of their competitors' bids and what their prospects of success are, further increases transparency and may even reduce corruption within the process [44].

The use of e-procurement may also reduce administration costs, the number of procurement staff required, and improve communication through swifter access to information the online availability of tender documents and information which can be updated regularly and promptly. Governments will further be able to publicly identify its regular suppliers thereby promoting transparency whilst at the same time promoting value for money in contracting with reliable suppliers. At the same time, competition can still be maintained by using various suppliers. The main benefit of introducing e-procurement recorded by the World Bank has been a marked upturn in transparency and competition [44].

6. Disadvantages of E-procurement

With e-procurement being a relatively new form of procuring goods and services, it has been up against a number of challenges, notwithstanding the proven benefits of using electronic means in procurement. In a 2004 UK wide study, Wong and Sloan in [48] found that only 48% of respondents indicated that they were able to conduct e-procurement effectively. The complete list of barriers ranked by this study was:

- Uncertain about the legal status of e-procurement;
- Organisational culture;
- Senior / executive management support;
- Lack of sufficient ICT infrastructure;

- ICT systems too expensive;
- Lack of technical skills and knowledge;
- Lack of qualified and proficient e-procurement personnel;
- Relationship challenges with suppliers when providing e-procurement;
- Possible security challenges during transactions;
- Interoperability issues; and
- No value for money [48].

Progress within the traditional procurement function has been slow and many governments have battled to gain traction, with varying levels of resistance to change [48] and an unwillingness to transfer to e-procurement systems, especially when structural change in business processes are necessary. In addition, the level of technological awareness of e-procurement system users to be, so-called "tech-savvy", is still lacking, requiring increased efforts to raise awareness of procurement-enabling technologies, and opportunities raised by these disruptive innovations [40]. In addition, ICT failure (whether that be a failure in the infrastructure or a failure in the equipment) during the procurement process is a reality; a lack of technical expertise knowledge and access to information technology are limitations to upcoming contractors. A lack of legal certainty underpinning e-procurement; and a failure by management to provide adequate infrastructure, are some disadvantages of e-procurement. The benefits of e-procurement can only be realised if the processes are properly improved and not simply by automating the existing methods of working [44].

A further possible threat is an increase in collusion where there are only a small number of suppliers of the required product or service. A lack of legal certainty underpinning e-procurement may be a further barrier to the successful implementation of e-procurement [48]. The benefits of e-procurement can only be realised through, amongst others, in-depth training of all stakeholders and adoption of an "electronic attitude" by suppliers. The use of mathematical formula's during evaluation can be problematic when public procurement is used as a tool to also achieve secondary objectives such as quality and socio-economic considerations. Different mathematical formulas must be developed for all types of procurement [44].

7. Conclusion

Although there are successes to be found where e-procurement has provided governments with the opportunity to improve their public procurement processes in terms of value for money, transparency, and integrity. Globally governments are only slowly starting to embrace the fourth industrial revolution, with technological advances being incorporated in order to survive in today's uncertainties. Public sector procurement entities are hard-pressed to accept the technology innovations to curb unnecessary costs and strengthen transparency measures to ensure that ethics are upheld throughout all the business processes [47]. In order to maximise the benefits of e-procurement, preparatory actions such as increased awareness of all stakeholders is essential as well as migration of all information pertaining to tender processes in order to successfully implement and maintain trust in an electronic procurement system. More importantly, senior public management support will be crucial in establishing and maintaining a trustworthy, transparent, and well-governed e-procurement system, supported by the necessary infrastructure and well-trained staff [44]. Faced with ever-increasing pressures on state funds to be spent for the greater good of all, the benefits of e-procurement will prove to outshine the disadvantages associated with a movement to technology. Less corruption, undue influence of tender awarding processes, and the reduction of other unethical behavioural risks are the advantages which will outweigh the disadvantages of the initial expenditure on technology, training of staff, and migration of all processes to an e-procurement environment. E-procurement provides the strategic foundation towards, transparent and cost-effective collaboration throughout the entire public procurement lifecycle [48].

Conflict of interest

The authors declare no conflict of interest.

Author details

David Fourie[1]* and Cornel Malan[2]

1 School of Public Management and Administration, Faculty of Economics and Management Sciences, University of Pretoria, Pretoria, South Africa

2 Independent Industry Advisor, Lyttelton, South Africa

*Address all correspondence to: prof.djfourie@up.ac.za

References

[1] OECD. Productivity in public procurement. A case study of Finland: measuring the efficiency and effectiveness of public procurement. Organisation for Economic Co-operation and Development (OECD). 2019. Available online: https://www.oecd.org/gov/public-procurement/publications/productivity-public-procurement.pdf. [Accessed : 2021-01-19]

[2] OECD. Understanding public sector productivity . Organisation for Economic Co-operation and Development (OECD). 2017. Available online: https://one.oecd.org/document/GOV/PGC[2017]10/en/pdf. [Accessed : 2021-01-18]

[3] Albassam, BA. A model for assessing the efficiency of government expenditure. Taylor & Francis Online, 2020. Available online at : https://www.tandfonline.com/doi/full/10.1080/23322039.2020.1823065 [Accessed : 2021-01-19]

[4] Aigheyisi, OS. Public Procurement, governance and Economic growth: some policy recommendations for Africa's growth and development., International Journal of Development and Management Review [INJODEMAR], Vol. 10. 2015. p. 110-124. Available on line: 120966-Article Text-332756-1-10-20150817.pdf. [Accessed : 2021-01-18]

[5] European Commission. Public Procurement. European Commission, 2019. Available online: https://ec.europa.eu/trade/policy/accessing-markets/public-procurement/ [Accessed : 2021-01-19]

[6] World Bank Blogs. How large is public procurement? [Online] 2020. https://blogs.worldbank.org/developmenttalk/how-large-public-procurement. [Cited: 2021-01-19]

[7] ITC. Empowering women through public procurement. International Trade Centre (ITC). 2014. Available online: https://www.tralac.org/images/docs/6257/itc-women-procurement-guide.pdf [accessed on 19 January 2021]

[8] Trybus, M. Supporting social considerations through public procurement: a legal perspective. Journal Article Forthcoming, 2016. Available online: SSRN-id2854205.pdf [Accessed : 2021-01-19]

[9] Nzitunga, J B. Strengthening the effectiveness of the procurement function in the public sector. Journal of Logistics Management, Vol. 8[2]. 2019, p. 25-34. Available online: 10.5923.j.logistics.20190802.01.pdf. [Accessed : 2021-01-19]

[10] Rejeb , A. Sűle , E and Keogh, J G. Exploring new technologies in procurement. Transport & Logistics: the International Journal, Vol. 18. 2018,p. 76-86.

[11] Fourie, David and Malan, Cornel. Public Procurement in the South African Economy. Sustainability. Vol. 12, 2020. p. 2-23. Available online: https://doi.org/10.3390/su12208692 [Accessed : 2021-01-19]

[12] Marye, T. Assessment of ethical practises in public procurement. The case of Woldia City Administration. (Unpublished master's thesis). Munich : Grin Verlag. 2015. Available online: https://www.grin.com/document/310681. [Accessed : 2021-01-18]

[13] OECD. Organisation for Economic Co-operation and Development - What is public procurement? [Online] 2020. https://www.oecd.org/governance/ethics/public-procurement.htm. [Cited: 2021-01-05]

[14] OECD. Public Procurement in Germany . [Online] 2019. https://www.oecd-ilibrary.org/governance/public-procurement-in-germany_1db30826-en. [Cited: 2021-01-18]

[15] McCue, CP. Prier, E. and Swanson, D. Five dilemmas in public procurement., Journal of public procurement, Vol. 15, 2015. p. 177-207. Available online: https://pdfs.semanticscholar.org/1f58/2e2ac641b666a84a8e1fbcbeed9dbe8198c1.pdf. [Accessed : 2021-01-20]

[16] OECD. Preventing Corruption in Public Procurement. Organisation for Economic Co-operation and Development. 2016. Paris, France

[17] UN. The United Nations - Procurement Practitioner's Handbook. [Online] 2012. https://www.ungm.org/Areas/Public/pph/ch03s01.html#sect_31 [Cited: 2021-01-20]

[18] OECD. Organisation for Economic Co-operation and Development - Public Procurement Toolbox. [Online] 2019. https://www.oecd.org/governance/procurement/toolbox/about/terminology/ [Cited: 2021-01-21]

[19] Bolton, P. The committee system for competitive bids in local government. Southern African Legal Information Institute, Vol. PER/PELJ 12, 2009, p20. Available online: http://www.saflii.org/za/journals/PER/2009/8.html. [Accessed : 2021-01-21]

[20] Arrowsmith, S. EU public procurement law: an introduction. [Online] 2010. https://www.nottingham.ac.uk/pprg/documentsarchive/asialinkmaterials/eupublicprocurementlawintroduction.pdf. [Cited: 2021-01-21]

[21] OECD. Organisation for Economic Co-operation and Development- Principles for integrity in public procurement. [Online] 2009. https://www.oecd.org/gov/ethics/48994520.pdf. [Cited: 2021-01-25]

[22] SIGMA. Support for Improvement in Governance and Management- Tender evaluation and contract award. [Online] 2011. http://www.sigmaweb.org/publications/47450401.pdf. [Cited: 2021-01-21]

[23] Khan, N. Public procurement fundamentals: lessons from and for the field. 2018. United Kingdom : Emerald Publishing Limited

[24] Ambe, IM. Public Procurement Trends and developments in South Africa., Research Journal of Business and Management, Vol. 3, 2016, p. 277-290. Pretoria : University of South Africa.

[25] Bolton, P R. Public Procurement as a Tool to Drive Innovation in South Africa., PER / PELJ, Vol. 19. 2016. p. 2-25. Available online: http://dx.doi.org/10.17159/1727-3781/2016/v19n0a1286. [Accessed : 2021-01-21]

[26] English Oxford Living Dictionaries. Definition - Ethics. www.en.oxford-dictionaries.com/definition/ethics. [Online] n.d. [Cited: 2021-01-19]

[27] Mle , TR. Professional and ethical conduct in the public sector. Africa's Public Service Delivery and Performance Review, Vol. 1, 2012. p. 27.

[28] Nagiah, K. The code of conduct of the South African Public Service compared with International Guidelines. (Unpublished Master's Dissertation). 2012 : University of Pretoria, South Africa.

[29] Fourie, D. Ethics in municipal supply chain management in South Africa. Local Economy, 2018, Available online: https://journals.sagepub.com/doi/abs/10.1177/0269094218809598. [Accessed : 2021-01-19]

[30] Makube, Tebogo. The Importance of using different methods of analysis in dealing with the challenges of collusive tendering and other forms of corruption in the South African Public Procurement System. African Public Procurement Law Journal, Vol. 6, 2019. p. 43-68. Available online : https://applj.journals.

ac.za/pub/article/view/26. [Accessed : 2021-01-19]

[31] Soreide, T. Corruption in public procurement - Causes, consequences and cures. 2002. Available online: https://www.cmi.no/publications/file/843-corruption-in-public-procurement-causes.pdf. [Accessed : 2021-01-12]

[32] Patras, MV. Critical phases in the process of awarding public procurement contracts. Romania case study., Journal of Public Administration, Finance and Law, 2016, Issue 8/2016 Available online: http://www.jopafl.com/uploads/issue8/critical_phases_in_the_process_of_awarding_public_procurement_contracts_romania_case_study.pdf. [Accessed : 2021-01-23]

[33] World Economic Forum. How governments can leverage policy and blockchain technology to stunt public corruption. [Online] 15 January 2020. https://www.weforum.org/agenda/2020/06/governments-leverage-blockchain-public-procurement-corruption/ [Cited: 2021-01-13]

[34] McCue, CP. Prier, E and Swanson, D. Five dilemmas in public procurement. Journal of Public Procurement, Vol. 15, Issue 2. 2015. p. 177-207. Available online: https://pdfs.semanticscholar.org/1f58/2e2ac641b666a84a8e1fbcbeed9dbe8198c1.pdf. [Accessed : 2021-01-23]

[35] Sarawa, DI and Mas'ud, A. Strategic public procurement regulatory compliance model with mediating effect of ethical behaviour. Heliyon. Vol. 6, Issue 1. 2020. p. 1-3. ISSN: 2405-8440.

[36] Mazibuko, GP. Analysis of the Administration of Procurement Practices in the South Africa Public Sector. 2018. Pretoria : University of Pretoria

[37] Nurmandi, A. and Kim, S. Making e-procurement work in a decentralized procurement system., International Journal of Public Sector Management, Vol. 28. Issue 3. 2015 p. 198-220. Available online: https://www.emerald.com/insight/content/doi/10.1108/IJPSM-03-2015-0035/full/pdf?title=making-e-procurement-work-in-a-decentralized-procurement-system-a-comparison-of-three-indonesian-cities. [Accessed : 2021-01-23]

[38] Neupane, A; Soar, J; Vaidya, K; and Yong, J. Role of Public E-Procurement Technology to reduce corruption in Government Procurement. International Public Procurement Conference, August 17-19, 2012. Seattle, Washington : Researchgate.

[39] Agwot, K. R. Public procurement in developing countries: objectives, principles and required professional skills., Public Policy and Administration, Vol. 6. Issue 8. 2016. p. 20-29. Available online: C:\Users\dalee\Desktop\Technological approach to ensure ethical procurement management\[PDF]Public Procurement in Developing Countries Objectives, Principles and Required Professional Skills.mht. [Accessed : 2021-01-21]

[40] Rejeb, A. Sűle, E. and Keogh, JG. Exploring new technologies in procurement. Transport & Logistics: the International Journal, Vol. 18. Issue 45. 2018. p. 46-86. ISSN 2406-1069

[41] Ndolo, J. and Njagi, E. Role of ethics in procurement process effectiveness in water sector in Kenya. International Journal of Supply Chain Management, Vol. 3, Issue 3. 2014. p. 129-133. Available online: https://www.researchgate.net/publication/281590064_Role_of_Ethics_in_Procurement_Process_Effectiveness_in_the_Water_Sector_in_Kenya_A_Case_Study_of_EWASCO_Embu_County [Accessed : 2021-01-25]

[42] Mathu, KM. The information technology role in supplier-customer information-sharing in the supply chain management of South African small and medium-sized enterprises. South

African Journal of Economic and Management Science, Vol. 22[1], a2256. 2019. p. 2-8. Available online at : https://sajems.org/index.php/sajems/article/view/2256. [Accessed : 2021-01-18]

[43] OECD. Organisation for Economic Co-operation and Development - Compendium of good practices for integrity in public procurement. [Online] 2014. https://www.oecd.org/gov/public-procurement/compendium-for-good-practices-for-integrity-in-public-procurement.pdf.[Cited: 2021-01-24]

[44] Allison, A. The Use of E-procurement in South African Public Procurement Law: Challenges and Prospects. 2018, Law, Democracy & Development, Vol. 22, p. 39-47. Available online: http://dx.doi.org/10.4314/ldd.v22i1.4. [Accessed : 2021-01-22]

[45] Nicoletti, B. The Future: Procurement 4.0. Agile Procurement. Cham : Palgrave Macmillan, 2018. p. 189-230. Available online : https://doi.org/10.1007/978-3-319-61085-6_8 [Accessed : 2021-01-26]

[46] Kayonde, E.N. Preferential procurement rules versus national treatment provisions in Investment Promotion and Protection Agreements. ZPPA Presentation with EU Trade Councillors : s.n., 2015.

[47] Manyathi, Sakhile. The Influence Of Procurement Practices On Service Delivery: An Analysis Of Government Legislation – Service Delivery Relationships In South Africa. 2019. Stellenbosch : University of Stellenbosch

[48] Eadie, R. Perera, S. and Heaney, G. Drivers and barriers to public sector e-procurement within Northern Ireland's construction industry., ITcon, Vol. 12. 2007 p. 103-120. Available online : https://itcon.org/papers/2007_6.content.07965.pdf [Accessed : 2021-01-26]

Chapter 3

The Desirability of a Future Integrated Reporting in the Study of Social and Innovative Practices

María-Victoria López-Pérez, Lourdes Arco-Castro, Jesús-Mauricio Flórez-Parra and Sara Rodríguez-Gómez

Abstract

Corporate social responsibility (CSR) has been considered the materialisation of ethics in organisations. CSR practices reflect companies' non-financial aspects, such as social and environmental issues. The proposal of an integrated report that jointly presents financial and non-financial issues would provide a global view of business activity, which will allow for analysis of the relationships and interactions among financial and non-financial resources involved in value creation, including human, social and relational, natural, and intellectual capital. If these resources are related, a report that integrates all of them would facilitate analysis. In this research, the relationship between innovation and CSR is studied. Environmental commitment may be a source of innovation (in the process of production and types of products) and involves social, relational, and intellectual capital changes. Innovation has previously been analysed from a perspective of competitiveness, necessitating a change of approach towards stakeholders that could allow us to reach a conceptual understanding of these relations. The research is empirically verified by studying a sample of 590 firm-years across 118 European companies that are leaders in sustainability, in the five-year period of 2011–2015. The results obtained show that CSR is a benchmark for addressing innovation and justifies the interest in an integrated reporting model that provides a global view of business.

Keywords: Integrated Report, Corporate Social Responsibility, Innovation, Stakeholders, Sustainability

1. Introduction

The concepts of corporate social responsibility (CSR) and business ethics have been used interchangeably in the existing literature [1]. While ethics is the set of principles and values that guides business behaviour, CSR is the set of socially and environmentally responsible practices of the company [2]. In recent decades, CSR has become an element to be integrated into the core of a business to allow the creation of value beyond the economic and ensure company longer term economic social viability of the company [3, 4].

Recent literature states that CSR could be oriented towards the search for value creation in terms of innovation [5, 6]. Numerous papers have suggested that CSR and innovation are related [7–10], but this relationship is not solidly proven and more research on the issue is necessary [9]. Some of this research states that their relationship increases operational efficiency by using cleaner technologies [11]. Another part of the research points out that CSR (if it were properly embedded across an organization) could improve performance through the development of innovative practices, processes, and products that enhance a company's competitive advantage through differentiation and cost saving strategies [12, 13]. Finally, other research, unrelated to competitive and operational aspects, shows that CSR linked to stakeholder management drives innovation in response to stakeholder demands, by improving companies' social performance [14–16]. Our research is included in this last group of studies and aims to analyse whether innovation is the consequence of CSR combined with the demands of company stakeholders.

One of the difficulties presented by the analysis of the relationship between CSR and innovation is the framework in which to place this object of study. The intuitive idea is clear: innovation and the concern for sustainability must be related and promote value in companies. However, possibly the biggest problem so far has been the lack of a business model that covers both concepts. The development of integrated reporting could provide a framework for studying these elements. The proposal of this type of report would be considered essential, as it would link the different responsibilities that companies assume. But the relationship between the different resources should first be demonstrated to justify the necessity of an integrated report [17]. Europe and the International Accounting Standards Board (IASB) are studying the suitability of making this type of report mandatory.

Integrated reporting provides an appropriate framework for CSR because in value creation, companies employ different types of capital, including natural capital, human capital, and social and relational capital [18] that, in the literature, have been grouped under the term CSR, which includes the responsibility assumed by the company in relation to these resources. CSR is a part of sustainability [19]. According to the Brundtland Report (1987), sustainable development is development that meets the needs of the present without compromising the needs of future generations [20]. Sustainability rests on three pillars: economic growth, environmental balance, and social responsibility. CSR can be defined as a company's responsibility for its impact on society [21], which leads to the integration into its corporate strategy of social, environmental, ethical, and human rights, and consumer concerns and commitments [21]. Integrated reporting is based on an accounting model that considers the responsible use of different resources to ensure sustainability or long-term value creation. The report must take into account the effort and investments made in CSR. In an integrated report, the elements that compose intellectual capital are knowledge-based intangibles. Among them are intellectual property (patents, licences, rights of exploitation and use of symbols, etc.) and organisational capital (tacit knowledge, systems, procedure and protocols). The element that may be most closely related to CSR is industrial property, and so this will be the focus of this research. Industrial property is the result of a process of innovation. Innovation can be considered as a process of discovery and development that gives rise to new products and production processes [22, 23]. Innovation is the application of knowledge to gain new knowledge that may be disruptive or incremental [24]. Companies are currently making great efforts in the field of innovation; in fact, it can be considered an inevitable step for any company that wants to grow, maintain, or create competitive advantages and/or access to new markets [5, 6]. Its importance for the survival and success of companies is widely accepted in the literature.

The first objective of this work is to deepen the theoretical framework around the relationship and interaction between CSR and innovation, setting stakeholder orientation (as opposed to the usual orientation to the market) as the study's approach to sustainability [14, 25]. This will serve to justify the interest of considering all these elements in a single document: an integrated report that is currently being considered for promotion by international organisations.

The second objective is to study the relationship between CSR and intellectual capital, specifically intellectual property. The results will allow us to verify contrast the integration of CSR and stakeholder orientation into the core business as a means of fostering innovation in companies. This justifies the importance that the holistic approach of integrated reporting will have in the study of value creation. The aim is to find that a company's social and relational capital creates intellectual value. The previous research into integrated reporting has mainly focused on the analysis of its adoption and its extension, but qualitative research on the possibilities of this type of reporting is scarce [26]. This highlights the relevance of our research, as the findings on the relationship between non-financial elements will be highly relevant in deciding whether to make integrated reporting that offers a global view of companies a mandatory requirement.

To achieve these objectives, the study is focused on Europe, because of the interest and effort of the European Commission to promote the development and disclosure of financial and non-financial information as well as the fact that CSR programs have different content according to the geographical environment in which they are implemented [27, 28]. The study is carried out on a wide sample; a CSR measure that considers all dimensions is taken and uses panel data that allows for control of unobserved heterogeneity, giving robustness to the model.

The work is structured as follows: after reviewing the relevant literature, we present the theoretical framework and propose the hypotheses to be compared. Following this, we describe the methodology used and present and discuss the results and findings. Finally, conclusions are drawn.

2. Background review and hypotheses presented

2.1 Integrated Reporting

In recent years, we have seen a growing trend in companies to consider multi-dimensional reporting that reflects the different elements involved in the development of business activity [29]. These reports integrate social, environmental, financial, and corporate governance information into a single document, the most widespread of which is the so-called integrated report, which aims to provide a synthesised and holistic view of organisations and their actions [17, 30].

In 2010, the International Integrated Reporting Council (IIRC) was formed with the participation of the main professional bodies and global accounting regulators - International Accounting Standards Board (IASB), Financial Accounting Standards Board (FASB), International Federation of Accountants (IFAC), and International Organization of Securities Commissions (IOSCO)- and other public bodies, the "Big Four" audit companies (Deloitte, EY, KPMG and PwC), leading multinationals, and representatives of institutions promoting social and environmental accounting [31]. The IIRC published the conceptual framework for integrated reporting, identifying a set of fundamental concepts and basic principles and contents for integrating sustainability into corporate objectives and reports [18, 32, 33].

Integrated reporting is based on two basic ideas. First, that a company's results involve the participation of resources of a varied nature, some of which are internal, and so controlled or owned by the company, and others that are external to the company, such as natural resources (water, air, land, flora) or those generated by society (social cohesion, effective governments, infrastructures, educational systems). Both types of resource are present in value creation [17, 31, 34]. These elements have to be considered in an integrated report to the extent that the company is accountable for the management of the resources used.

The second idea refers to sustainability. Value creation is not only to be understood in a financial sense; it also implies that there is a balance between the variations experienced by the various capitals, both internal and external, in the development of the business activity. The decreases in some of the capitals, mainly the external ones, and should be properly justified. This leads to the need for a reporting model that goes beyond the financial model and comprehensively considers the resources that allow the creation of value. Such a report can be also used as a management tool.

The analysis of sustainability requires combined consideration of the ecological, economic, and social effects that occur in the development of business activity and that can affect the availability of resources in the future. It refers to the responsibility of organisations to integrate economic, environmental, and social aspects into operations and business strategy [21] to assure the viability of the enterprise in the medium and long term [26, 35, 36]. It is assumed that there are relationships between all these elements that companies use and that it is necessary to ensure that they are real. Therefore, these dimensions should not be considered in isolation but should take into account their synergies and interrelations [19], which lead to medium and long term value creation [18, 32]. The underlying idea is that the combined effect of the different capitals is greater than the individual contribution of each of them. This leads to the idea that there must be a relationship between these capitals, which can and should be analysed prior to the study of their contribution to value creation.

2.2 Theoretical framework

In relation to CSR disclosure, several theoretical frameworks have been used [37, 38]. The most widely used of these are: institutional theory, which is appropriate when it is necessary to analyse the incidence of normative, institutional, and cultural contexts, etc. [39, 40]; the legitimacy theory, which is based on the existence of a social contract or licence to operate between companies and society [38, 41]; and stakeholder theory, which states that the responsibilities of companies towards society have significantly expanded [42]. Stakeholder theory is the most used, useful, and dominant theory to explain sustainability practices and is applicable in the context of this research [43]. The non-financial aspect included in integrated reporting involves consideration of the participation of different resources and stakeholders in value creation and sustainability [30]. Integrated reporting requires CSR to be part of the corporate and core business strategy. According to stakeholder theory, suppliers of factors understood in a broad sense, i.e. the five types of capital indicated above, are involved and associated with the organisation and cooperate to ensure the survival and continuity of the firm [31]. An integrated report should respond to the needs and interests of key stakeholders (investors, consumers, employees, suppliers and community) [18].

The role of companies and their commitments to society, employees, other stakeholders, and the environment is changing [44, 45]. Stakeholder theory requires linking the behaviour of a company with the effects on its stakeholders. In this context, the company must take into account the stakeholders' interests in products, behaviours, and programs developed by the entity. Stakeholders are an essential element in the success or failure of an entity [46]. An integrated report should respond to the interests of the groups involved and to some extent implies the application of accountability for the use of financial and non-financial resources, such as intellectual, social or relational capitals.

Stakeholder theory requires that companies balance the legitimate but sometimes conflicting interests of stakeholders [46, 47]. This requires considering their management and providing them with information [46]. A company's future success is linked to its consideration of and response to stakeholder expectations [48, 49].

Companies have to manage the stakeholders that directly or indirectly collaborate with the entity to achieve its objectives [45, 50]. This aspect of stakeholder theory fits into the framework of integrated reporting. Stakeholder theory has been widely considered in the literature as a solid justification for both social and environmental disclosure practices and for corporate governance mechanisms. In this sense, it is also applicable in the combined consideration of all these elements in a single report [26].

2.3 The orientation towards stakeholders and innovation

Although stakeholder theory is widely accepted in relation to CSR and innovation, most research focuses on competitiveness, obtaining competitive advantages, [5] and analysing the effects of programs on performance indicators [51, 52]. Emphasis has been placed on the effect of these practices on the market or investors. However, in the last few years, we have seen the model evolve towards a broader vision, where CSR and innovative practices could generate value beyond economic and commercial benefits [5].

Recent literature states that CSR could be oriented towards the search for value creation in terms of innovation for the company and society [5, 6]. Some of this research shows that CSR drives innovation in response to stakeholder demands [14, 15]. These works have focused on a more ethical vision of the relationship between CSR and innovation. Innovation in itself can generate social benefits, such as the generation of more economic products, the creation of new jobs [53], and the development of more sustainable business models [11, 54]. In this sense, entities could establish innovative practices that respond to the demands and expectations of stakeholders to ensure the creation of value.

Stakeholders often have unused or untapped knowledge that complements a company's internal knowledge and is valuable in achieving the goal of sustainable value creation [12]. The importance given to stakeholders in the elaboration of CSR programs is evidenced by entities' establishment of relational networks and new channels of communication to obtain information about stakeholder demands, expectations, and perceptions. Attention to suggestions made by environmental agencies, research institutes, community, consumers, employees, and investors and, where appropriate, integration into CSR programs can help strengthen stakeholder relations. Engaging with stakeholders allows companies to identify innovation opportunities [55]. The active participation of stakeholders helps in the detection stage and favours efficiency in the development of new proposals avoiding the development of ideas that are not in demand in the market. Subsequently, the consideration of different interests in management makes it possible to create situations of mutual benefit for companies and society [3, 56]. The interests of the different stakeholders can be aligned with the concept of shared value, the company survives in the market through innovation and the companies meet different needs of its stakeholders. The concept of shared value underlies the integrated report.

According to the above, there is a positive relationship between orientation to stakeholders and the development of innovative practices. As a result of companies' focus on stakeholders in CSR programs, CSR is expected to have a greater impact on innovation. The following hypotheses are proposed:

Hypothesis 1. A company's orientation towards stakeholders encourages innovation.

Hypothesis 2. A company's stakeholder orientation positively moderates the effect of CSR on innovation.

3. Material and methods

3.1 Size and characteristics of the sample

CSR can be defined in many ways and measured using many different approaches. In the present study, we focus on a sample of European firms that form part of the Dow Jones Sustainability Index (DJSI). The firms in this index are leaders in the field of CSR. To qualify for incorporation into the index, they must conform to very demanding CSR guidelines (based on economic, social, and environmental indicators that will be included in the integrated report) and are rated according to these guidelines by the Sustainability Index of the Sustainable Asset Management (SAM) Group [57]. This score was utilised in the present study. The indexed companies develop practices that go beyond legal requirements and respond to ethical values and commitments demanded by society.

The period 2011–2015 is examined, obtaining an initial sample of 176 European firms that formed part of the DJSI. From the total number of European companies included in this index, we removed 41 that were dedicated to financial and insurance activities and a further 17 that have not been in the index throughout the entire period analysed. Accordingly, the final sample consisted of 118 firms.

The sample is distributed by country and sector as shown in **Tables 1** and 2.

3.2 Variables selected

Innovation is the dependent variable. Innovation can be measured through output indicators (product and process innovations, patents) [58] or input indicators (R&D expenditure). Integrated reporting chooses to measure innovation through industrial property, i.e. innovation is measured by the number of patents registered (PAT) [59]. Patents have the advantage of being an objective element and a

Country	Frequency (number of companies)	Percent
Belgium	1	.8
Denmark	4	3.4
Finland	3	2.5
France	15	12.7
Germany	14	11.9
Hungary	1	.8
Ireland	1	.8
Italy	5	.8
Netherlands	8	4.2
Norway	4	6.8
Portugal	1	3.4
Spain	9	7.6
Sweden	9	7.6
Switzerland	12	10.2
United Kingdom	31	26.3
Total	118	100

Table 1.
Countries in the sample.

Country	SIC CODE	Frequency (number of companies)	Percent
Mining, construction	100–1979	15	12.7
Manufacturing	2000–3999	56	47.5
Transportations, Communications, Electric, Gas and Sanitary service	4000–4999	23	19.5
Wholesale Trade	5000–5199	12	10.2
Retail	5200–5999	12	10.2
Total		118	100

Table 2.
Industries in the sample.

measure of the results obtained from R&D activities [60, 61]. Moreover, it provides a measure of a firm's current technological capacities, efficiency, and potential future profits from R&D [62]. In addition, it constitutes a mechanism that favours the appropriation of the benefits obtained from innovation [63] and the capacity to create added value [64].

In Europe, the adoption of patent protection tends to increase as firms grow [65]. Patents have been considered the most believable proxy of innovation [65]. The patents corresponding to each of the firms in our study were compiled from information disclosed by the Spanish Patent and Trademark Office (OEPM) for each of the years considered in this study.

Stakeholder orientation and CSR were taken as the independent variables.

On the one hand, stakeholder orientation (STAKE) is measured through the existence of mechanisms and channels of communication that aid the active participation and collaboration of stakeholders and provide possibilities for interaction [66]. Specifically, the characteristics of interactivity, the existence of forums/chats, and the existence of web 2.0 technologies (websites and social networking sites that allow users to share information and interact with each other), online surveys, and information sheets are analysed [67–69]. To this end, the websites of the companies selected each year from those making up the sample are reviewed.

On the other hand, CSR is a multidimensional construct that takes into account various dimensions and aspects- social, environmental and economics- [25, 70, 71] in accordance with the aims of this study. Many researchers use a single CSR measure, such as environmental performance, philanthropic contributions, corporate policies, revealed misdeeds, transparency, or investment in health and safety [72]; but this only considers one aspect of CSR. Among the multidimensional measures most commonly used are Kinder Lydenberg Domini's *Socrates* database [7, 12, 73] and the Fortune magazine database. In recent years, stock indices have been set up with components including sustainability. In the present paper, CSR is measured using the DJSI score in the period referred to above for each of the companies included in our sample [57].

We also included a moderating variable"STAKE*CSR", to reflect the joint effect of the two variables. This will allow us to observe whether a company's stakeholder orientation positively moderates the effect of corporate social responsibility on innovation.

Finally, control variables were included referring to the firm's size, risk, and the industry sector in which it was active [9]. Size was measured using the logarithm of total asset (ASSETS) [71, 74]; the industrial sector (IND) was measured in accordance with the standard industrial classification code, thus creating a 5-block group [60, 75, 76]; and risk (RISK) was measured by the firm's debt/asset ratio [75].

3.3 Methodology

Panel data econometric analysis was used to test the hypotheses proposed in this paper. Specifically, a random effect model (GLS regression) was utilised after applying the Breusch-Pagan and Hausman tests. Panel data provide consistent data from samples for which repeated observations of cross-section data are available; in the present case, this refers to firms over a period of various consecutive years. Thus, no information is lost. In addition, the use of panel data makes it possible to control unobserved heterogeneity, which would otherwise bias the results [77, 78]. Therefore we eliminate the possibility of aggregation bias that can arise when using mean data for the variables, in time series models. The use of random effects has advantages over fixed effects, such as the problem of incidental parameters, being appropriate for random samples of large populations or allowing the treatment of omitted factors [79].

Panel data allow for the introduction of dynamic elements into the model. All this is why this analysis has been used in the recent literature on CSR and innovation [79, 80]. To test the hypotheses, the following model was considered:

$$PAT = b_1 + b_2\,STAKE + b_3\,RSC + b_4\,STAKE * RSC + b_5 ASSETS + b_6 IND + b_7 RISK + e.$$

4. Results and discussion

Tables 3 and **4** show the descriptive statistics and correlations, and **Table 5** shows the results obtained after applying the linear model to the panel data.

With respect to the indices of correlation (**Table 4**), there is evidently a positive, significant association between stakeholder orientation and innovation. The existence of communication channels to obtain information on stakeholder demands is a source of ideas that could allow the company to develop its capacity for innovation [55]. On the other hand, innovation in itself can generate social benefits, which justifies stakeholders' demand for it [11, 53, 54]. In this respect, stakeholders could be promoting innovation practices in the company. This highlights the idea of shared value that underlies integrated reporting. The management of different capitals generates mutual benefits [3, 56]. In this case, the management of social and relational resources would develop entrepreneurial innovativeness.

	M	Std.Dev.	Min	Max
Dependent Variable				
Patents (PAT)	19.92	28.36	0	233
Independent variables				
Stakeholders Orientation (STAKE)	2.04	0.41	1.83	4.16
CSR	66.40	14.05	33	91
Control variables				
Assets	7.36	7.63	3.39	8.41
Risk	0.61	0.19	0.03	1.56
Industry (IND)	3.55	1.41	1	5

Table 3.
Descriptive statistics of the variables used (N = 118).

	1	2	3	4	5
1. PAT	1.00				
2. STAKE	0.19**	1.00			
3. CSR	0.21**	0.38**	1.00		
4. ASSETS	−0.02	0.01	0.02	1.00	
5. RISK	−0.22**	0.02	−0.11*	−0.00	1.00
6. INDUSTRY	0.12*	0.02	0.04	0.14*	0.21**

$^{*}p < .05$; $^{**}p < .01$.
Pearson's correlation coefficient.

Table 4.
Correlations between dependent, independent and control variables.

Dependent variable		**PAT**	
	Model 1 (n = 118)	**Model 2 (n = 118)**	**Model 3 (n = 118)**
Constant	5.09*** (5.51)	15.32*** (3.50)	1.30*** (5.97)
CSR	0.16*** (0.06)		0.22*** (0.07)
Stake		0.06** (0.02)	0.33** (0.17)
Stake*CSR			0.01† (0.00)
Risk	−7.07 (4.45)	−9.52** (4.32)	−8.36** (4.38)
Ind	−0.19 (0.74)	0.05 (0.70)	−0.04** (0.72)
Assets	0.00 (0.00)	0.00 (0.00)	0.00 (0.00)
Adjusted R Square	0.281	0.286	0.299
Wald	10.91 (4)	9.57 (4)	16.94 (6)
Probability	0.027	0.048	0.009
Rho	0.172	0.133	0.141

$^{\dagger}p < .10$; $^{**}p < .05$; $^{***}p < 0.01$.
Standard errors appear in parenthesis.

Table 5.
Regression analysis.

The analysis of correlations shows a relationship between CSR and patents [68, 73, 75]. Greater effort in the field of CSR is reflected as a higher level of innovation (measured by the number of patents). The results of innovation are associated with CSR [9, 75]. In selecting differentiation strategies, some companies decide on CSR, and this strategy requires innovation activities [53, 81]. CSR constitutes an organisational resource that incorporates various policies, among which is that of innovation. CSR provides a framework for developing innovation [82, 83]. When planning innovation, companies must take into account the priorities determined by CSR. Thus, the latter may be utilised as a means of directing innovation [84]. They may also respond to the fact that the adoption of a CSR strategy requires changes be made to production processes or new products be introduced, ones that are more environment-friendly – and these considerations are relevant in the innovation process [85].

The positive correlation between CSR and stakeholder orientation shows that socially responsible companies address the demands and interests of their stakeholders and integrate them into their CSR practices [79]. In this sense, socially responsible companies could integrate stakeholders' social and environmental demands into their innovation and development strategies. On the other hand, the correlation suggests that there is a stakeholder demand for corporate social responsibility practices [50, 86].

CSR practices are negatively and significantly associated with levels of debt, which could mean that companies with a lower level of financial risk are more likely to adopt CSR practices. The financial structure of the company determines its capacity for innovation. The greater capacity to access sources of financing allows a greater inflow of financial resources that can be applied to various strategies, including innovation [79]. Finally, the results show a significant correlation between the industry sector and innovation, which indicates that the sector in question is a significant factor with respect to the introduction of technological change [23]. Due to the need for mechanisation of their processes, certain industries have seen their capacity for innovation fostered.

Model 1 of **Table 5** shows that CSR has a positive and significant effect on innovation. The companies analysed are leaders in sustainable and socially responsible practices, so it seems logical to think that they use CSR to generate intangible assets, such as industrial and intellectual property [82, 87]. From a management standpoint, the adoption of sustainability practices has a positive impact on value creation [13].

Model 2 describes the relationship between stakeholder orientation and innovation, and shows that the examined companies' innovation efforts are positively associated with stakeholder orientation. These companies are implementing innovation as a means to respond to the interests of stakeholders. In this sense, we accept Hypothesis 1. In recent decades there has been an increasing demand for more environmentally friendly processes, practices, products, and services. This social demand has triggered a wave of innovation in companies that are more oriented towards stakeholders and society in general [79]. Thus, the results show that more stakeholder-oriented companies generate more patents. Relationships, networks, and collaborative mechanisms between the company and the groups of interest are effective mechanisms for capturing new social and environmental needs and developing innovation capabilities to address them [45, 50]. **Table 6** shows the description of the variable stakeholder orientation.

Model 3 shows the possible moderating effect of stakeholder orientation on the relationship between CSR and innovation. The variables in this model are of greater statistical significance, and thus we conclude that the impact of CSR is enhanced by stakeholder orientation. Stakeholder demands encourage the effect of CSR on innovation. Stakeholder orientation should be included in the business strategy to boost research and development in the company [55, 79]. In accordance with the above results, Hypothesis 2 is accepted. Thus, empirical research shows that in order to enhance the effect of corporate social responsibility on innovation, it is necessary for the company to know the demands and interests of its stakeholders, and communication channels are a good means of achieving that objective. In this sense, the results suggest that CSR generates intellectual capital when it generates social value by fostering relationships with stakeholders [14, 15]. The different business capitals are related as the integrated report points out.

Stakeholder orientation		$STAKEH = \sum_{i=1}^{M} gi$
Concept	**Items**	**Score**
1. Characteristics of interactivity	a. An e-mail address other than that of the webmaster is provided for requests for information or explanations. b. Personal contacts with the persons responsible at the university for the information provided are provided on the website. c. The website has a mailing list to update information to users of the information applying this service.	0/0.33 on the basis of the absence-presence of each item
2. Forums/chat	a. General forums b. CSR-related forums	0.5 if the online forum/chat used allows for discussion of general issues 1 if there is a specific forum/chat used for the discussion of CSR issues
3. Web2.0 technology	a. An e-mail address other than that of the webmaster is provided for requests for information or explanations. b. Personal contacts with the persons responsible at the university for the information provided are provided on the website. c. The website has a mailing list to update information to information users who apply this service.	0/0.33 on the basis of the absence-presence of each item
4. Online surveys	a. General content forums b. CSR-related forums	0.5 if the online forum/chat used allows the discussion of general topics and 1 if there is a specific forum/chat used for the discussion of CSR topics.
5. Newsletter	a. General content forums b. CSR-related forums	0.5 if the online forum/chat used allows the discussion of general topics and 1 if there is a specific forum/chat used for the discussion of CSR topics.

Source: Own elaboration based on previous literature [67–69].

Table 6.
Description of the variable stakeholder orientation.

5. Conclusions

The determinant relationships between innovation, CSR (social and environmental practices), and stakeholder orientation show that there is a real link among them, and as a result, it would be necessary for the company to adopt a holistic vision that takes into account different capitals (natural, human, social, and relational) to ensure the creation of value and the generation of assets [18]. In this sense, our study shows that CSR and stakeholder orientation promote intellectual capital, industrial property, in leading European companies in sustainability, and an integrated report that includes all the resources will allow for better management of them.

CSR constitutes a framework incorporating various policies, one of which is innovation. Some of the policies on innovation are related to those concerning CSR, which indicates that companies may seek to differentiate themselves from their competition by means of their CSR strategy. Innovation is a difficult factor to control but, as shown in the results, its links to CSR provide a suitable context for appropriate implementation. A finding of the research is that innovation policies are aimed at goals that are in accordance with CSR practices [87].

Moreover, taking CSR as a variable mediated by stakeholder orientation, we conclude that there is a joint effect on innovation. The integration of these two strategies generates a greater number of patents. The research shows that stakeholder orientation may require changes to production processes or products, and hence a re-orientation of innovation policy may be required [55]. An additional finding is that the resulting attention to the social and environmental demands of stakeholders could encourage more sustainable practices and processes, which could generate shared value [3, 48].

It would be interesting in subsequent research, to examine the extent to which CSR practices require innovations involving radical change or inventions, or whether the innovations made are mere developments of existing technology. In addition, it could be interesting to analyse concepts such as eco-innovation. Furthermore, as risk constitutes a significant factor, a further study should be made of the effect of a firm's ownership structure on the CSR strategies adopted and on its innovation policy. Future research could study the impact of different stakeholders on innovation policies (such as employees and consumers) and analyse the possible impact of corporate governance, which could improve the analysis.

Conflict of interest

The authors declare that there is no conflict of interest.

Annex.Description of the "stakeholder orientation" variable

Author details

María-Victoria López-Pérez*, Lourdes Arco-Castro, Jesús-Mauricio Flórez-Parra and Sara Rodríguez-Gómez
Faculty of Economy and Business, University of Granada, Campus de Cartuja, Granada, Spain

*Address all correspondence to: mvlopez@ugr.es

References

[1] Fombrun C, Shanley, M: What's in a Name? Reputation Building and Corporate Strategy. Academy of Management Journal. 1990;33:233-258.

[2] Epstein E M: The corporate social policy process: Beyond business ethics, corporate responsibility, and corporate social responsiveness. California Management Review. 1987;29:99-114.

[3] Porter, M E, Kramer M R: Shared value. Harvard Business Review. 2011;1:5-18.

[4] Wójcik P: The business case for corporate social responsibility: A literature overview and integrative framework. Journal of Management and Business Administration. 2018;26: 121-148.

[5] Martinez-Conesa I, Soto-Acosta P, Palacios-Manzano M: Corporate social responsibility and its effect on innovation and firm performance: An empirical research in SMEs. Journal of cleaner production. 2017; 142:2374-2383.

[6] Zhou H, Wang Q, Zhao X: Corporate social responsibility and innovation: A comparative study. Industrial Management & Data Systems, 2020; 5:863-882.

[7] Chakrabarty S, Wang L: The long-term sustenance of sustainability practices in MNCs: A dynamic capabilities perspective of the role of R&D and internationalization. Journal of Business Ethics. 2012; 110:205-217.

[8] McWilliams A, Siegel D, Wright P: Corporate social responsibility: Strategic implications. Journal of Management Studies. 2006;43:1-18.

[9] Surroca J, Tribó J A, Waddock S: Corporate responsibility and financial performance: The role of intangible resources. Strategic Management Journal. 2010;31:463-490.

[10] Wagner M: Corporate social performance and innovation with high social benefits: A quantitative analysis. Journal of Business Ethics. 2010;94: 581-594.

[11] Geissdoerfer M, Bocken N M, Hultink E J: Design thinking to enhance the sustainable business modelling process–A workshop based on a value mapping process. Journal of Cleaner Production. 2016;135:1218-1232.

[12] Luo X, Du S: Exploring the relationship between corporate social responsibility and firm innovation. Marketing Letters. 2015;26:703-714.

[13] Rexhepi G, Kurtishi S, Bexheti G: Corporate social responsibility (CSR) and innovation the drivers of business growth. Procedia-Social and Behavioral Sciences. 2013;75:532-541.

[14] Li X: The effectiveness of internal control and innovation performance: An intermediary effect based on corporate social responsibility. Plos one. 2020;15: e0234506.

[15] Weng H H, Chen, Ja-Shen, Chen, Pei-Ching: Effects of Green Innovation on Environmental and Corporate Performance: A Stakeholder Perspective. Sustainability. 2015;7:4997-5026.

[16] Waheed A, Zhang Q, Rashid Y, Tahir M S, Zafar M W: Impact of green manufacturing on consumer ecological behavior: Stakeholder engagement through green production and innovation. Sustainable Development. 2020;28:1395-1403.

[17] Burke J J, Clark C E: The business case for integrated reporting. Insights from leading practitioners, regulators, and academics. Business Horizons. 2016;59:273-283.

[18] IIRC, (2013). The International <IR> Framework. London(United

Kingdom): International Integrated Reporting Council.

[19] Ashrafi M, Adams M, Walker T R, Magnan G: How corporate social responsibility can be integrated into corporate sustainability: A theoretical review of their relationships. International Journal of Sustainable Development & World Ecology. 2018;25:672-682.

[20] WCED, S. W. S. (1987). World commission on environment and development. Our common future, 17, 1-91.

[21] Unión Europea (2011). Comunicación de la comisión al parlamento europeo, al consejo, al comité económico y social europeo y al comité de las regiones. Estrategia renovada de la UE para 2011-2014 sobre la responsabilidad social de las empresas (COM2011).

[22] OCDE (2005). Manual de Oslo: Guía para la recogida e interpretación de datos sobre innovación. Recuperado de http://www.conacyt.gob. sv/ Indicadores% 20Sector% 20Academcio/ Manual_de_Oslo.

[23] Van Kleef J A G, Roome N J: Developing capabilities and competence for sustainable business management as innovation: a research agenda. Journal of Cleaner Production. 2007;15:38-51.

[24] Zortea-Johnston E, Darroch J, Matear S: Business orientations and innovation in small and medium sized enterprises. International Entrepreneurship and Management Journal. 2012;8: 145-164.

[25] Chang K, Kim I, Li Y: The heterogeneous impact of corporate social responsibility activities that target different stakeholders. Journal of Business Ethics. 2014;125:211-234.

[26] Hahn R, Kühnen M: Determinants of sustainability reporting: A review of results, trends, theory, and opportunities in an expanding field of research. Journal of Cleaner Production. 2013;59:5-21.

[27] Lopez M, Garcia A, Rodriguez L: Sustainable development and corporate performance: A study based on the Dow Jones Sustainability Index. Journal of Business Ethics. 2007;75;285-300.

[28] Maon F, Swaen V, Lindgreen A: One vision, different paths: An investigation of corporate social responsibility initiatives in Europe. Journal of Business Ethics. 2017;143:405-422.

[29] Havlová K: What integrated reporting changed: The case study of early adopters. Procedia Economics and Finance, 2015;34:231-237.

[30] Perego P, Kennedy S, Whiteman G: A lot of icing but little cake? Taking integrated reporting forward. Journal of Cleaner Production. 2016;136:53-64.

[31] Flower J: The international integrated reporting council: A story of failure. Critical Perspectives on Accounting. 2015;27: 1-17.

[32] IIRC. (2015). The international <IR> Framework. London(United Kingdom): International Integrated Reporting Council.

[33] James M L: Sustainability and integrated reporting: A case exploring issues, benefits and challenges. Journal of the International Academy for Case Studies. 2013;19:89.

[34] Camilleri M A: Theoretical insights on integrated reporting: The inclusion of non-financial capitals in corporate disclosures. Corporate Communications: An International Journal. 2018;23:567-581.

[35] Brogi M, Lagasio V: Environmental, social, and governance and company profitability: Are financial intermediaries different?. Corporate Social

Responsibility and Environmental Management. 2019;26:576-587.

[36] Chowdhury R H, Choi S, Ennis S, Chung D: Which Dimension of Corporate Social Responsibility is a Value Driver in the Oil and Gas Industry?. Canadian Journal of Administrative Sciences/Revue Canadienne des Sciences de l'Administration. 2019;36:260-272.

[37] Hooghiemstra R: Corporate communication and impression management–new perspectives. Why companies engage in corporate social reporting. Journal of Business Ethics. 2000;27:55-68.

[38] Matuszak Ł, Różańska E: Online corporate social responsibility (CSR) disclosure in the banking industry: Evidence from Poland. Social Responsibility Journal. 2019;16:1191-1214.

[39] Frias-Aceituno J V, Rodríguez-Ariza L, García-Sánchez I M: Is integrated reporting determined by a country's legal system? An exploratory study. Journal of Cleaner Production. 2013;44:45-55.

[40] Oware K M, Mallikarjunappa T: CSR expenditure, mandatory CSR reporting and financzial performance of listed firms in India: an institutional theory perspective. Meditari Accountancy Research.DOI:10.1108/MEDAR-05-2020-0896.

[41] Patten D M: Seeking legitimacy. Sustainability Accounting Management and Policy Journal. 2020;11:1009-1021.

[42] Masoud N, Vij A: The effect of mandatory CSR disclosure on firms: Empirical evidence from UAE. International Journal of Sustainable Engineering. 2020. https://doi.org/10.1080/19397038.2020.1821405

[43] Hörisch J, Schaltegger S, Freeman R E: Integrating stakeholder theory and sustainability accounting: A conceptual synthesis. Journal of Cleaner Production. 2020;275:124097.

[44] Carroll A B, Shabana K M: The business case for corporate social responsibility: A review of concepts, research and practice. International Journal of Management Reviews. 2010;12:85-105.

[45] Yasser Q R, Al Mamun A, Ahmed I: Corporate social responsibility and gender diversity: Insights from Asia Pacific. Corporate Social Responsibility and Environmental Management. 2017; 24;210-221.

[46] Freeman R E: Strategic management: A stakeholder approach. Advances in strategic management. 1983;1:31-60.

[47] Wood D J: Corporate social performance revisited. Academy of Management Review. 1991;16:691-718.

[48] Porter M E, Kramer M R The competitive advantage of corporate philanthropy. Harvard Business Review. 2002;80:56-68.

[49] Cantrell J E, Kyriazis E, Noble G: Developing CSR giving as a dynamic capability for salient stakeholder management. Journal of Business Ethics. 2015;130:403-421.

[50] Madsen P M, Rodgers Z J: Looking good by doing good: The antecedents and consequences of stakeholder attention to corporate disaster relief. Strategic Management Journal. 2015;36:776-794.

[51] Briones Peñalver A J, Bernal Conesa J A, de Nieves Nieto C: Analysis of corporate social responsibility in Spanish agribusiness and its influence on innovation and performance. Corporate Social Responsibility and Environmental Management. 2018;25:182-193.

[52] Mishra D R: Post-innovation CSR performance and firm value. Journal of Business Ethics. 2017;140;285-306.

[53] Kalkanci B, Rahmani M, Toktay L B: The role of inclusive innovation in promoting social sustainability. Production and Operations Management. 2019;28:2960-2982.

[54] Hu B, Zhang T, Yan S: How corporate social responsibility influences business model innovation: The mediating role of organizational legitimacy. Sustainability. 2020;12:2667.

[55] Herrera-Baltasar M E: Creating competitive advantage by institutionalizing corporate social innovation. Journal of Business Research. 2015;68: 1468-1474.

[56] Jones T M, Felp, W: Shareholder wealth maximization and social welfare: A utilitarian critique. Business Ethics Quarterly. 2013;23:207-238.

[57] Zink K J: Stakeholder orientation and corporate social responsibility as a precondition for sustainability. Total Quality Management and Business Excellence. 2005;16:1041-1052.

[58] Henderson R, Cockburn I: Scale, scope, and spillovers: the determinants of research productivity in drug discovery. Rand Journal of Economics. 1996;27: 32-59.

[59] Grant R M, Contemporary strategy analysis: Concepts, techniques, application. Basil Blackwell, Massachusetts; 1992.

[60] Ernst H: Patent applications and subsequent changes on performance: evidence from time-series cross-section analyses on the firm level. Research Policy. 2001;30:143-157.

[61] Lin B, Chen J: Corporate technology portfolios and R&D performance measures: A study of technology intensive firms. R&D Management. 2005;35:157-170.

[62] Hirschey M, Richardson V J, Scholz S: Value relevance of nonfinancial information: The case of patent data. Review of Quantitative Finance and Accounting. 2001;17:223-235.

[63] Patel P, Pavitt K, editors. (Ed.), Handbook of the economics of innovation and technological change. Patterns of technological activity: Their measurement and interpretation. In P. Stoneman, 14-51. Oxford: Blackwell Publishers; 1995.

[64] Nakamura L: Economics and the New Economy: The invisible hand meets creative destruction. Federal Reserve Bank of Philadelphia Business Review. 2000;15-30.

[65] Watanabe C, Tsuji Y S, Riffy-Brown C: Patent statistics: Deciphering a 'real' versus a 'pseudo' proxy of innovation. Technovation. 2001;21:783-790

[66] Du S, Bhattacharya C B, Sen S: Maximizing business returns to corporate social responsibility (CSR): The role of CSR communication. International Journal of Management Reviews. 2010;12: 8-19.

[67] Fang, Z: E-government in the digital era: Concept, practice, and development. International Journal of the Computer. The Internet and Management. 2002; 10;2: 1-22.

[68] Garde Sánchez R, Flórez-Parra J M, López-Pérez M V, López-Hernández A M: Corporate governance and disclosure of information on Corporate Social Responsibility: An analysis of the Top 200 universities in the Shanghai ranking. Sustainability. 2020; 12: 1549-1571.

[69] Holzer M, Manoharan A: Global trends in municipal e-government: An online assessment of worldwide municipal web portals. Foundations of E-government, Computer Society of India. 2007: 177-178.

[70] Griffin J J, Mahon J F: The corporate social performance and corporate financial performance debate:

Twenty-five years of incomparable research. Business and Society. 1997;36:5-31.

[71] McWilliams A, Siegel D: Corporate social responsibility and financial performance: correlation or mis-specification?. Strategic Management Journal. 2000;21:603-609.

[72] Peloza J: The challenge of measuring financial impacts from investments in corporate social performance. Journal of Management. 2009;35:1518-1541.

[73] Padgett R C, Galan J I: The effect of R&D intensity on corporate social responsibility. Journal of Business Ethics. 2010;93: 407-418.

[74] Ullmann A A: Data in search of a theory: a critical examination of the relationships among social performance, social disclosure, and economic performance of U.S. firms. Academy of Management Review. 1985;10:540-557.

[75] Hull C E, Rothenberg S: Firm performance: the interactions of corporate social performance with innovation and industry differentiation. Strategic Management Journal. 2008;29: 781-789.

[76] Waddock S A, Graves S B: The corporate social performance-financial performance link. Strategic Management Journal. 1997;18:303-319.

[77] Bouquet C, Deutsche Y: The impact of corporate social performance on a firm's multinationality. Journal of Business Ethics. 2008;80:755-769.

[78] Martínez-Ferrero J, Ruiz-Cano D, García-Sánchez I M: The causal link between sustainable disclosure and information asymmetry: The moderating role of the stakeholder protection context. Corporate Social Responsibility and Environmental Management. 2016; 23:319-332.

[79] García-Piqueres G, García-Ramos R: Is the corporate social responsibility–innovation link homogeneous?: Looking for sustainable innovation in the Spanish context. Corporate Social Responsibility and Environmental Management. 2020;27:803-814.

[80] Yu F, Shi Y, Wang T: R&D investment and Chinese manufacturing SMEs' corporate social responsibility: The moderating role of regional innovative milieu. Journal of Cleaner Production. 2020;258:120840.

[81] Zeimers G, Anagnostopoulos C, Zintz T, Willem A: Organisational learning for corporate social responsibility in sport organisations. European Sport Management Quarterly. 2019;19:80-101.

[82] Halkos G, Skouloudis A: Corporate social responsibility and innovative capacity: Intersection in a macro-level perspective. Journal of Cleaner Production. 2018;182:291-300.

[83] Russo M V, Fouts P A: A resource-based perspective on corporate environmental performance and profitability. Academy of Management Journal. 1997;4:534-559.

[84] Hockerts K, Morsing M: A literature review on corporate social responsibility in the innovation process. Report of the Center for Corporate Social Responsibility Copenhagen Business School (CBS). 2008:1-28.

[85] Siegel D, Vitaliano D: An empirical analysis of the strategic use of corporate social responsibility. Journal of Economics and Management Strategy. 2007; 17:773-792.

[86] Zerbini F: CSR initiatives as market signals: A review and research agenda. Journal of Business Ethics. 2017;146:1-23.

[87] Guiral A: Corporate social performance, innovation intensity and financial performance: Evidence from lending decisions. Behavioral Research in Accounting. 2012;24:65-85.

Section 2

Ethical Best Practice

Chapter 4

Modern Privacy Threats and Privacy Preservation Techniques in Data Analytics

Ram Mohan Rao P, S. Murali Krishna and AP Siva Kumar

Abstract

Today we are living in a digital rich and technology driven world where extremely large amounts of data get generated every hour in the public domain, which also includes personal data. Applications like social media, e-commerce, smartphone apps, etc. collect a lot of personal data which can harm individual privacy if leaked, and hence ethical code of conduct is required to ensure data privacy. Some of the privacy threats include Digital profiling, cyberstalking, recommendation systems, etc. leading to the disclosure of sensitive data and sharing of data without the consent of the data owner. Data Privacy has gained significant importance in the recent times and it is evident from the privacy legislation passed in more than 100 countries. Firms dealing with data-sensitive applications need to abide by the privacy legislation of respective territorial regions. To overcome these privacy challenges by incorporating privacy regulations, we have designed guidelines for application development, incorporating key features of privacy regulations along with the implementation strategies which will help in developing data-sensitive applications which can offer strong and coherent privacy protection of personal data.

Keywords: Data privacy, ethical code of conduct, privacy legislations, privacy preservation, design strategies

1. Introduction

Broad use of smart phones, e-commerce, social media, Internet and Communication Technologies (ICT) has transformed our lives. Though digitization facilitates our work, it is prone to privacy vulnerabilities. The key privacy threats include surveillance, disclosure, targeted advertisements [1], identity theft, information disclosure without consent, personal abuse through cyber stalking [2], studying emotions and mood of the people by accessing profile pictures, tweets, likes and comments to find emotionally weak, people who are lonely and trap them using various cyber-attacks like ransom ware, sexual abuse etc. [3]. Firms dealing with data sensitive applications need to follow certain ethical code of conduct to ensure privacy and guard the users from various digital assaults.

Digital data include variety of personal data like transactional data, location data, electronic medical records, e-commerce data, insurance data, photos and videos, opinions and views etc. All these data items are personal and sensitive data and should not be disclosed without the consent of the data owner. Privacy breach can occur at various stages of data processing as described in **Figure 1**.

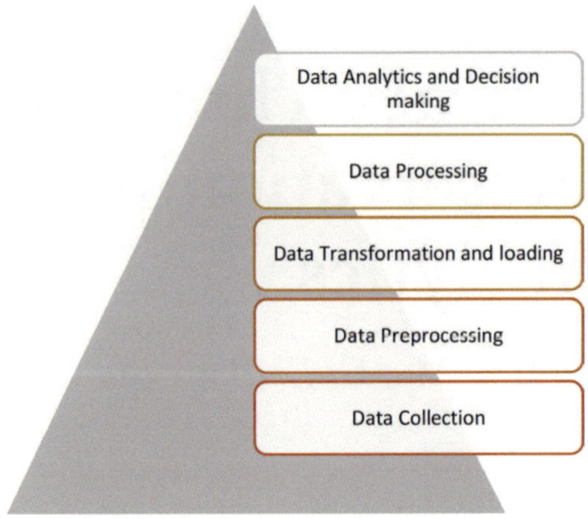

Figure 1.
Data processing stages.

Data breach can occur at any stage of data processing with different type of people operating at various levels. The top-level management should ensure that no data breach occur at any stage for which there is a need to have a policy in place and ethical code of conduct for all employees of the firm. However, policies alone are not sufficient; there must be regulatory body to ascertain the policies are implemented. Apart from this, the individuals are also contributing to data leakage by inappropriate use of social media and smart phones. Hence there are three entities responsible to ensure privacy preservation (**Figure 2**).

Governments and regulatory bodies are more responsible than others because Governments can impose privacy regulations and make sure the data holder or data collecting firms abide by them. Data holders are also equally responsible because data is with them and they can share the data with third parties without the knowledge of the data owner. By inappropriate use of social media applications like Facebook, Instagram, etc. users are also uploading personal data into the public domain which leads to privacy threats. With the consistent escalation of privacy threats and their grave consequences, awareness among users has also increased and in turn increased the demand for privacy preservation, which eventually led to the creation of privacy laws and regulations in many countries. The most prominent among them are GDPR (General Data Protection Regulation) of the European Union and the Personal Data Protection bill of India. Some of the applications along with its privacy risk are listed in **Table 1**.

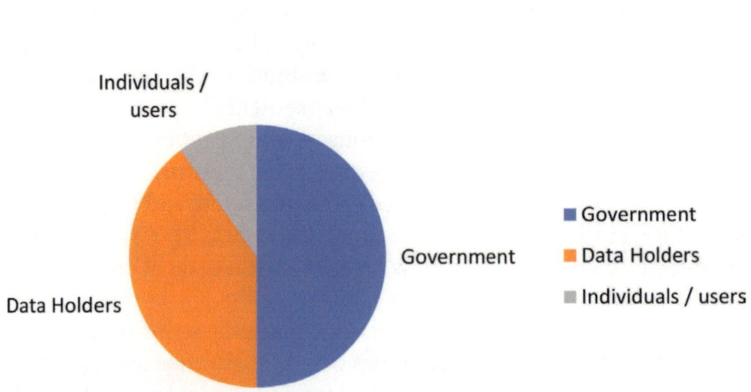

Figure 2.
Entities responsible for privacy preservation.

DOI: http://dx.doi.org/10.5772/intechopen.99160

S.no.	Application type	Privacy risk involved
1	Smart phone apps.	Information theft, Intrusion
2	e-Commerce sites	Inference attacks, Disclosure
3	Social media	Cyber stalking, Ransom ware
4	Data capturing systems like banking, hospitals, insurance, government portals etc.	Disclosure, Discrimination

Table 1.
Application vs. privacy risk.

2. Ethical code of conduct through privacy legislations

Incredible amount of digital data is generated by virtue of various applications and technologies. This digital data will also contain personal and sensitive data of an individual which must be confidential, secure and private. Data privacy is the responsibility of the firms that capture the data and ensure no privacy breach in any stage of data processing. Hence there is a need for ethical code of conduct for privacy preservation of personal and sensitive data in private and public domains. **Figure 3** depicts few important practices of ethical code of conduct in data processing.

There is a need for privacy legislations because of modern privacy threats and also to ensure ethical practices being followed by data holders. Some of the modern privacy threats are:

a. Profiling

b. Social media privacy threats

c. Privacy hazards in image analytics

2.1 Digital profiling

Digital Profiling is the automated processing of person specific data to evaluate certain attributes relating to a person, particularly to analyze and predict

Figure 3.
Ethical practices in data processing.

individual's economic situation, buying habits, health, preferences, interests, behavior, etc. Digital Profiling also influences group privacy wherein an individual may be a member of one or more groups [3]. Digital Profiling is widely used in direct digital marketing businesses. Profiling without consent of the individual is a privacy breach. Cookies are a piece of data stored in the browsers, when users browse and transact. Cookies are used in auto profiling and cookies can be read without user's consent. Google has recently announced to end support for third party cookies in its Chrome browser which will make it very difficult for the digital marketing companies to build a user profile [4]. Article 22 of GDPR facilitates the right to the individual that, no automated data processing including profiling is allowed without consent from the user.

2.2 Privacy threats in social media applications

Social media platforms are highly vulnerable to stalking attacks. One of the common stalking techniques involves an online mob of anonymous self-organized groups to target individuals, causing defamation, threats of violence, and technology-based attacks. Social media are used to build trust between the perpetrator and the victim. Perpetrator is a person who may carry out a harmful, illegal and immoral act. When victim transmits confidential data including pictures and videos, the perpetrator can intercept, steal confidential data and abuses them for blackmail purposes [5]. Social media firms are also responsible to identify user with such malicious intentions, block them and initiate appropriate legal actions as per the law.

2.3 Privacy hazards in image analytics

Image data analytics is widely used in health care, social media, and e-commerce applications. In social media large numbers of images are uploaded every day. An image is worth more than a thousand words and hence it may reveal the emotional state of a person also [6]. Some of the key privacy hazards in image data analytics include

a. Attempt to analyze emotional state of people and exploit them. Facebook and WhatsApp status updates can be studied using machine learning models and sentiment analysis can help analyze the social and emotional wellbeing of a person and in turn exploit them.

b. Disclosure of secret medication being taken by a person by virtue of promotional offers on medicine.

c. Another important privacy concern is identity theft, because copies of permanent account number (PAN) cards, passports and driving licenses are kept in digital form and shared. Insurance, banking firms and third parties will extract lot of sensitive data which is a serious privacy hazard [7, 8].

d. Medical imaging deals with visual representation of internal structure of organs and tissues. Medical imaging may lead to leakage of personal and sensitive medical data of a person [9].

Ethical code of conduct in digital data processing is required for privacy preserving data processing and it is possible only through stringent implementation of privacy regulations. More than hundred countries have passed legislations to protect individual privacy and GDPR of European Union was the pioneer. All other privacy policies and regulations are framed based on GDPR. The key features of GDPR and other privacy regulations are.

1. **Right to forget or erase data:** personal data gets uploaded in many digital applications. For example, people upload certain private photos and videos, buy certain products online and if users wish, the data or transactional records can be removed from their databases.

2. **Users consent before sharing the data:** Data holder's share and exchange data for real time insights, but in many applications the data owner is not aware of it. It is required to take the consent of the data owner before sharing.

3. **No surveillance without consent:** many applications will monitor their user's behavior including location, device type etc. Data profiling companies and digital advertisement companies do surveillance without consent from the user and most of the users are not even aware of surveillance. It is now mandatory for all firms to take user consent for surveillance, in the countries where privacy legislations are in force.

4. **Right to restrict the data processing:** many data intensive applications, process data without prior consent of the data owner. It is mandatory to take prior permission from the data owner to use data for further processing.

Failure to affirm the privacy compliance will attract serious consequences including huge amounts of fine and detriment to reputation. To ensure privacy preservation and abide by the local privacy regulations, firms are undergoing changes in their policies to incorporate privacy regulations. Following strategic changes were noticed in the year 2020 [10].

1. 30% of the businesses made changes in their cyber security models due to GDPR and 60% of them created new policies.

2. 15% firms offered extra training to staff on communications and GDPR

3. 11% firms have changed their firewalls and system configurations.

4. 6% created new contingency plans.

3. Application design strategies

Applications that cause serious privacy threats were listed in **Table 1**. For each application, design strategies and guidelines are provided, so that the applications cannot harm the privacy of the user.

3.1 Design guidelines for smart phone apps

It is a common practice that most of the users do not read the privacy policy and the network permissions which an app demands before installation. People ignore and will agree for all permissions the app demands which lead to serious privacy concerns. To ensure inherent privacy protection, smart phone apps must be designed with following features.

a. Seek only the minimum permissions for the app to be functional.

b. Do not collect any metadata including location, type of device, time etc.

c. No auto profiling of the user by any app is allowed.

d. Accept and abide the federal laws of the region or state pertaining to data access and sharing.

e. Design to ensure no access to any free Wi-Fi which is not registered by the user.

f. Do not transfer any data from the phone without consent from the user.

g. Privacy policy should not be a text document. Privacy policy should be an audio file played in the language opted by the user, ensure the user listens to it completely and finally accepts or rejects the privacy policy. Polling can be used to ensure user's attention. i18n (internationalization) applications are required and easy to develop with present open source technology frameworks to offer privacy policy as an audio file in the language opted by the user. i18n applications are the applications that offer multilingual user interface. It is the process of writing software so that it can support local languages and cultural settings.

i18n applications: The word i18n represents internationalization. In the word "internationalization", the number of characters between the first and last characters i.e. i and n are 18, hence the name i18n. Applications are said to be i18n applications when they support multilingual user interface. Applications read the request headers to know the language preferences of the user. For example if the user's language preference is Spanish, then the user interface will automatically reflect the content in Spanish. Generally i18n is applied to web applications. In web applications, http protocol is used for request and responses. When a http request is made, along with the request few request headers will also be sent to the web application. One of the request header is "*accept-language*" which contains the language which the user prefers to use. These language preferences can be changed by the users through browser settings. If the web application is i18n enabled then it will read the value of the *accept-language* header and display the user interface in the language mentioned by the user. Such applications are called i18n applications.

3.2 Design guidelines for e-commerce sites

e-commerce sites use recommendations to offer value added services to the customers. Recommendations are used as part of improved service. However, there is always a possibility of information disclosure. For example, a person wanted to buy some product for personal use. He/she wanted this to be confidential and by virtue of recommendations, he/she may see a pop up or alert showing a better offer on that product which is visible to the people sitting nearby and this will lead to discrimination and personal embarrassment. Based on the type of products bought, the gender of the person can also be inferred which is an unwanted disclosure. In order to ensure privacy protection, following features need to be incorporated in the design of the e-commerce sites in line with the privacy legislations.

1. Privacy Quotient (P_μ): Recommendations are used by ecommerce firms to provide value added services and best possible offers to the customers based on their buying habits and transaction history. Recommendation systems lead to

serious privacy concern which is not addressed by any ecommerce firm and the same is illustrated here. For example a person regularly bought some product online, related to personal care and does not want to disclose this to anyone. However, since it is a regular transaction the ecommerce firm would like to recommend the same product to him by offering decent discount on the product and the same will displayed on his screen when he/she logs into their account and it is a privacy breach if someone else sees the same. It can lead to discrimination of the person in the family or profession. To address this problem, we introduce the concept of privacy quotient. For every product the ecommerce firm should provide an option where in user can opt, whether this product and purchase is to be made private or not, thereby excluding it from any form of analytics or recommendations. If 40% buyers of a product opt for transaction privacy i.e. the product purchase is not to be used for recommendations, then the product must be considered as private and for all buyers of this product, the transaction must be made private. This percentage of transactions which decide the transaction privacy is called as privacy quotient (P_μ) [11].

2. No sharing of data without users' consent: No e-commerce site, must share customers data without consent. However, data can be shared with federal authorities for any investigation purpose.

3. Meta data: e-commerce sites tend to collect metadata including location, type of device used, IP address etc. without the permission and knowledge of the user. It has to be avoided.

3.3 Social media platform design issues

Social media has emerged as the most vulnerable platform of privacy abuse especially cyber stalking, ransom ware, sexual abuse etc. Important issues to be addressed in social media applications are

1. Identification of fake accounts and stringent mechanism of anomaly detection.

2. Deep neural networks can be used in identifying the private and sensitive information in the images uploaded by the user, remove them and store the modified image. User consent is mandatory. Users must be advised of privacy threats every time when they upload photos or videos.

3.4 Data capturing systems

Disclosure and discrimination are the common threats related to data capturing systems. Organizations like hospitals, banks, retail supply chain etc. collect a lot of person specific data while offering respective services. This data will be analyzed to gain deep insights and come up with better decisions and offer value added services.

1. As per the privacy regulations across many countries, it is recommended to use non-anonymized and model based solutions for privacy preservation.

2. Sensitive attributes must be tokenized before sharing with any other third party for analytics.

3. Quasi identifiers must be synthesized before sharing.

3.5 Data privacy officer

Every organization that deals with personal and sensitive data must employ a Data Protection Officer (DPO). DPO must be a technology expert with sound knowledge on privacy policies and regulations.

DPO is responsible for ethical code of conduct and implementation of privacy laws of the respective region or territory. Some of the key responsibilities of the DPO are

1. Provides complete security to the data.

2. Records all the activities performed on the data.

3. Seeks consent from the data owner, every time when the data is processed

4. Responds to the queries of the customers or data owners.

5. Ensures implementation of local privacy policies and federal laws.

6. Notification of privacy breach if any to the data owner

7. Impact assessment

4. Conclusions

As part of our work, we proposed few guidelines for application design which will support individual privacy in many data intensive applications in line with privacy legislations. These days more privacy violations and abuse are being reported in social media where people upload lot of personal photos and videos. Huge number of fake profiles were also reported who may indulge in activities like cyber stalking, ransom ware etc. There is a need for strong and coherent privacy preservation mechanism for social media applications and has enough scope for research especially employing deep learning models.

Acknowledgements

I thank Dr. G. Kiran Kumar for his valuable inputs in writing my first book chapter.

Conflict of interest

The authors declare no conflict of interest.

Author details

Ram Mohan Rao P[1*], S. Murali Krishna[2] and AP Siva Kumar[1]

1 Department of Computer Science and Engineering, JNTUA Ananthapuramu, Andhra Pradesh, India

2 Department of Computer Science and Engineering, Sri Venkateswara College of Engineering Tirupathi, Andhra Pradesh, India

*Address all correspondence to: rammohan04@gmail.com

References

[1] Ducange, Pietro, Riccardo Pecori, and Paolo Mezzina. "A glimpse on big data analytics in the framework of marketing strategies." Soft Computing 22.1 (2018): 325-342.

[2] Dhillon, Gurpreet, and Kane J. Smith. "Defining objectives for preventing cyberstalking." Journal of Business Ethics 157.1 (2019): 137-158.

[3] Mavriki, Paola, and Maria Karyda. "Automated data-driven profiling: threats for group privacy." Information & Computer Security (2019). https://doi.org/10.1108/ICS-04-2019-0048

[4] https://www.cnbc.com/2020/01/14/google-chrome-to-end-support-for-third-party-cookies-within-two-years.html

[5] March, Evita, et al. "Somebody that I (used to) know: Gender and dimensions of dark personality traits as predictors of intimate partner cyberstalking." Personality and Individual Differences 163 (2020): 110084. https://doi.org/10.1016/j.paid.2020.110084

[6] Chen, Lushi, et al. "Building a profile of subjective well-being for social media users." PloS one 12.11 (2017). https://doi.org/10.1371/journal.pone.0187278

[7] Yang, Jingjing, Jinzhao Wu, and Xiaojing Wang. "Convolutional neural network based on differential privacy in exponential attenuation mode for image classification." IET Image Processing (2020). https://doi.org/10.1049/iet-ipr.2020.0078

[8] Beaulieu-Jones, Brett K., et al. "Privacy-preserving generative deep neural networks support clinical data sharing." Circulation: Cardiovascular Quality and Outcomes 12.7 (2019): e005122. https://doi.org/10.1161/CIRCOUTCOMES.118.005122

[9] P. Wang, T. Chen and Z. Wang, "Research on privacy preserving data mining," Journal of Information Hiding and Privacy Protection, vol. 1, no.2, pp. 61-68, 2019.doi:10.32604/jihpp.2019.05943;

[10] https://www.nvtgroup.co.uk/keep-safe-this-data-privacy-day/

[11] Rao, P. Ram Mohan, S. Murali Krishna, and AP Siva Kumar. "Novel algorithm for efficient privacy preservation in data analytics." (2021).

Chapter 5

An Ontology for Standardising Trustworthy AI

Dave Lewis, David Filip and Harshvardhan J. Pandit

Abstract

Worldwide, there are a multiplicity of parallel activities being undertaken in developing international standards, regulations and individual organisational policies related to AI and its trustworthiness characteristics. The current lack of mappings between these activities presents the danger of a highly fragmented global landscape emerging in AI trustworthiness. This could present society, government and industry with competing standards, regulations and organisational practices that will then serve to undermine rather than build trust in AI. This chapter presents a simple ontology that can be used for checking the consistency and overlap of concepts from different standards, regulations and policies. The concepts in this ontology are grounded in an overview of AI standardisation currently being undertaken in ISO/IEC JTC 1/SC 42 and identifies its project to define an AI management system standard (AIMS or ISO/IEC WD 42001) as the starting point for establishing conceptual mapping between different initiatives. We propose a minimal, high level ontology for the support of conceptual mapping between different documents and show in the first instance how this can help map out the overlaps and gaps between and among SC 42 standards currently under development.

Keywords: Trustworthy AI, AI Ethics, AI Governance, AI Standards, Ontology

1. Introduction

To be effective, future regulation and organisational policy aimed at achieving trustworthy AI must be supported by some degree of standardisation in processes and technological interoperability. The rapid development of AI technologies and the growth of investment in AI applications presents a pacing problem, wherein the rapid change in characteristics of AI related to policy and regulatory issues outpaces ability of societies to legislate for or regulate the technology. At the same time, the multinational nature of the major commercial developers of AI, plus the expanding access to AI skills and computing resources means that standards must be agreed internationally to be of widespread use in supporting policy and regulations. While there has been an explosion in policy documents from national authorities, international organisations and the private sector on the ethical implications of AI, standards in this area have been slower to emerge. Understanding existing standardised ICT development and organisational management practices offer insight into the extent to which they may provide a basis for standardising practice in governing the development and use of more trustworthy and ethical AI. Standards Developing Organisations (SDOs) vary in their approach to addressing specific ethical issues.

The Institute of Electrical and Electronic Engineers (IEEE) global initiative on ethically aligned design for autonomous and intelligent systems has spawned the IEEE 7000 standards working group that places ethical issues at its heart [1]. This work was seeded from a set of principles defined in a comprehensive international export review on Ethically Aligned Design [2], which also highlighted the influence of classical ethics, professional ethics and different moral worldviews.

A different approach is taken by the ISO/IEC Joint Technical Committee 1 (JTC1). JTC1 which was established by the International Standardisation Organisation (ISO) and the International Electro-technical Commission (IEC) in 1987 to develop, maintain and promote standards in the fields of Information Technology (IT) and Information and Communications Technology (ICT). Expert contributions are made via national standards bodies and documents (over 3000 to date) are often used as technical interoperability and process guideline standards in national policies and international treaties, as well as being widely adopted by companies worldwide. Statements of relevance to UN Sustainable Development Goals and Social Responsibility Guidelines are an inherent part of all new standardisation projects proposed in JTC 1 [3]. AI standards are addressed together with big data technology standards by the JTC 1 subcommittee (SC) 42 which was first chartered in autumn 2017 and held its inaugural meeting in April 2018. As of the end of 2020 it has published six standards and has active projects addressing 23 others (https://www.iso.org/committee/6794475.html).

This chapter highlights the challenges facing companies and authorities worldwide in advancing from the growing body of work on ethical and trustworthy AI principles to a consensus on organisational practices that can deliver on these principles across the global marketplace for AI-based ICT. We review how SC 42 standardisation efforts benefit from building on established process standards in areas of management systems, IT governance, risk and system engineering. From this analysis, we identify a simple conceptual model that can be used to capture the semantic mapping between different SC 42 standards. An ontology is used as it allows a conceptual model to be defined that links together concepts via association into a network of concepts. This has the potential to establish an open ontology that can map between core concepts from standardisation and pre-standardisation deliverables in varied states development, formal approval, and international community consensus with concepts needed to address trustworthy AI. Such a network allows the definition of terms and concepts from different standards related documents to be interlinked and thereby the consistency of conceptual use between different can be analysed and improvements suggested. While this is not intended to replace the consistency checking that occurs naturally in the JTC1 standards development process, it does allow us to identify some mapping and comparisons between different forms of standard that have been applied to different areas of standardisation in SC 42. We conclude then by suggesting how this approach can be extended to enable similar comparisons with the use of concepts in documents being drafted by other SDO committees and by other bodies, including regulatory proposals, civic society policy proposals and guidelines developed by individual organisations.

2. Challenges of building international consensus on governing trustworthy AI

Since 2017 there has been an explosion in AI initiatives globally. As of February 2021, the Council of Europe's tracker (https://www.coe.int/en/web/

artificial-intelligence/national-initiatives) has identified over 450 such initiatives world wide, primarily from national authorities, international organisations and the private sector. The most frequently discussed address subjects include privacy, human rights, transparency, responsibility, trust, accountability, freedom, fairness and diversity. Influential works such as the IEEE EAD [2], the EU's High Level Expert Group on AI [4] and the OECD [5] often present these issues under the banner of ethical or trustworthy AI.

Scholars and think tanks have analysed this growing body of documents on ethical and trustworthy AI. One extensive survey identifies an apparent consensus on the importance of ethical principles of transparency, justice, non-maleficence, responsibility, and privacy, whereas other issues of sustainability, dignity, and solidarity in relation to labour impact and distribution garner far less attention across works [6]. Public authority works are identifying gaps in relation to the use of AI by governments and in weapon systems [7]. Private sector outputs have been criticised as instruments to reduce demand for government regulation [8], as potential barriers to new market entrants [9] and failing to address tensions between ethical and commercial imperatives within organisations [10]. A general criticism is a focus on individual rather than collective harms such as loss of social cohesion and harm to democratic systems [11]. The required progression from approaches that propose broad principles, to specific and verifiable practices that can be implemented by organisations and, where deemed necessary, regulated by legislation, implies a focus on governance and management of AI. Appropriate governance, management, and risk management measures can reinforce benefits and mitigate the ethical and societal risks of employing AI technology. Governance approaches can be characterised as [12]: market-based, resulting from value-chain partner pressures, including from consumers; self-organisation, based on an organisation's internal policies; self-regulation based on industry wide agreement on norms and practices; and co-regulation based on industry compliance with government regulation and legislation. There have been some proposals for possible regulatory structures including: new national [13] and international [14] co-regulatory bodies and internal (self-regulatory) ethics boards that may help organisations implement best practice [15, 16]. However, AI governance through co-regulation presents a number of major challenges [17]. These include: reaching stable consensus on what defines AI; widening access to AI skills and computing infrastructure obscuring developments from regulators; the diffusion of AI development over locations and jurisdictions globally; the emergence of impacts of an AI system only when assembled into a product or service; the opacity of modern subsymbolic machine learning methods and techniques, i.e. their unsuitability for clear human readable explanations; the potential for highly automated AI-driven systems to behave in unforeseeable ways that can escape the visibility or control of those responsible for them. More broadly, co-regulation is challenged by: the pacing problem, as AI technology develops faster than society's ability to legislate for it; international cooperation needed for common standards being impeded by AI's perceived role as a strategic economic or military resource; the perceived impediments of legislation to realising the competitive national economic and social benefits of AI; and the power asymmetry in AI capability being concentrated in digital platforms benefiting from network effects [9]. Over all types of works, a wide range of motivation have been identified [18], the incompatibility of some of which can further impede consensus on approaches to implementing trustworthy AI.

Nevertheless, there are multiple parallel standardisation activities ongoing internationally that are attempting to build some level of consensus, including the above-mentioned IEEE P7000 and ISO/IEC JTC 1/SC 42 activities and several national activities. This multiplicity of standards development may itself, however, contribute to inconsistencies and incompatibilities in how different organisations govern their AI activities. Reducing ambiguity in how different stakeholders in the AI value chain communicate with each other, and with society in general about their trustworthy AI practices is therefore critical to building trustworthiness of the resulting AI-based products and service. With both individual organisations developing their own AI policies and legislation for AI regulation starting to be considered in major jurisdictions such as the EU, there is a need to support the ongoing mapping of concepts between these different parallel activities so that harmful or expensive inconsistencies can be identified early and hopefully resolved.

The following requirements for semantic interoperability between concepts developed by different bodies can therefore be identified and are depicted in **Figure 1**:

1. Consistency between documents being developed within the same standards family.

2. Comparison between documents produced by parallel standards activities.

3. Mapping between standards and regulatory/legislative proposal to determine the extent to which a standard can address regulatory requirements (US "safe harbour" or EU's harmonised standards approach).

4. Analysing the degree of compliance between a standard and an organisation's AI policies and procedures, including their documentation outputs.

5. Assessing organisational policies against regulations.

6. Comparing different regulatory proposals internationally or comparing revisions to proposals in a single jurisdiction.

7. Comparing AI policies published by different organisations.

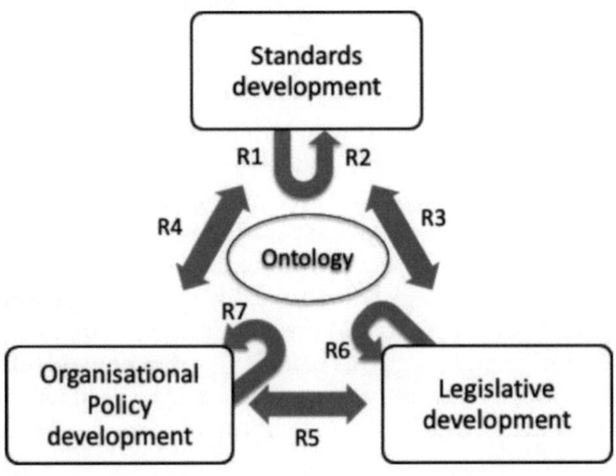

Figure 1.
Role of semantic interoperability between bodies involved in governance of trustworthy AI.

3. JTC1 standards related to trustworthy AI

Same as other ISO/IEC JTC 1 standardisation activities, SC 42 places a strong emphasis on ensuring consistency with existing process and interoperability standards as well as reuse of existing terms and concepts to provide industry with a consistent body of applicable standards. SC 42 is therefore addressing AI-related gaps within existing standards, including those for management systems, risk management, governance of IT in organisations, IT systems and software quality. Rather than addressing AI ethics directly as a normative issue, SC 42 addresses the broader issues of trustworthy AI, with a technical report that sets out some of the core concepts and issues for standardisation related to Trustworthy AI (ISO/IEC 24028:2020) [19]. In this report, trustworthiness is defined as the ability to meet stakeholders' expectations in a verifiable way. When applied to AI, trustworthiness can be attributed to services, products, technology, data and information as well as organisations when considering their governance and management. This view considers trustworthy AI as realisable as part of a broader set of engineering, management, and governance process standards that can be employed together by organisations involved in AI and that can support mechanisms for conformity assessment including 3rd party certification and external oversight.

The Trustworthiness Working Group (WG 3) within SC 42 has a strong pipeline of pre-standardisation and standardisation activities. The road mapping activities within the group are driven by gap analyses of prior art as well as current policy documents (including the IEEE EAD [2], HLEG [4], and OECD [5]). WG 3 builds on foundational terminology and high-level life cycle notions elaborated within SC 42/WG 1 foundational deliverables ISO/IEC CD 22989 [20] on AI Concepts and Terminology and ISO/IEC CD 23053 [21] on a Framework for Artificial Intelligence (AI) Systems Using Machine Learning. WG 3 primarily looks at Trustworthiness high level characteristics and addresses them through elaboration of new project proposals of either pre-standardisation informative deliverables, surveying the state of the art in an area (before proceeding to normative coverage at a later stage) or of normative deliverables.

The fully fledged normative deliverable type within the ISO/IEC ecosystem is an International Standard (IS), however, not many areas in AI are mature enough to be addressed in international standards. This includes non-normative technical reports on current approaches to addressing societal and ethical aspects of AI [22] and bias in AI [23]. Thus WG 3 currently works only on three IS deliverables.

ISO/IEC CD 23894 Information Technology — Artificial Intelligence — Risk Management [24] is a specialisation of ISO 31000 Risk Management [25]. This is an example of SC 42's respect for prior art and application of existing frameworks such as quality, risk management, or management systems framework in the newly standardised area of AI and ML.

Another IS deliverable within the group is ISO/IEC WD 25059 Software engineering — Systems and software Quality Requirements and Evaluation (SQuaRE) — Quality model for AI-based systems [26]. This IS is an extension to the influential Systems and software Quality Requirements and Evaluation (SQuaRE) series owned by JTC 1/SC 7. Quality and trustworthiness are in a sense competing paradigms as they are looking at similar sets of high-level characteristics such as robustness, reliability, safety, security, transparency, explainability etc. but the distinctive difference is in the need for quality stakeholders to take explicit part in actively defining quality requirements, while trustworthiness stakeholders do not have to explicitly state their expectations in order to influence objective trustworthiness criteria. At any rate, the SQuaRE4AI standard sets a quality model that profiles the traditional quality and trustworthiness top level characteristics and their sub-characteristics for other normative deliverables in the area that are aiming at setting method and process requirements and recommendations.

The third IS in the making in WG 3 is ISO/IEC WD 24029–2 Artificial intelligence (AI) — Assessment of the robustness of neural networks — Part 2: Methodology for the use of formal methods [27]. This series aims to address the technical robustness pillar of AI trustworthiness, Part 2 specifically by looking at formally provable robustness and performance related properties of neural networks. While machine learning and neural networks in particular are an extremely active R&D field, the formal mathematical theory in which neural networks are based is well academically researched and stable. Therefore, it is possible to benefit from known and provable properties of neural networks in current and upcoming industrial applications.

Technical Specification is a normative deliverable that has a less rigorous approval process, essentially there is only one round of national bodies approval for a TS compared to two distinct (and repeatable) stages for an IS approval. While it is easier to approve and publish a TS, a TS needs to be transformed into an IS or withdrawn 3 years after its publication. TS are sometimes called experimental standards. This type of deliverable is used in areas that are in urgent need of normative standardisation as demonstrated by industry or societal demand, while the area to be standardised is still in flux from the research and development point of view. This is why WG 3 decided to ask SC 42 national bodies to approve development of ISO/IEC NP TS 6254 Information technology — Artificial intelligence — Objectives and methods for explainability of ML models and AI systems.

To develop an understanding of how these trustworthy AI standards relate to policies and processes defined by individual organisations and emerging as regulations in different jurisdictions requires an understanding of other aspects of AI standardisation under development in the other working groups of SC 42.

Working group 1 (WG 1) addresses foundational standards including the above-referenced AI Concepts and Terminology [20], which aims to provide consistency across the use of terms and concepts in other SC 42 documents and a Framework for Artificial Intelligence (AI) Systems Using Machine Learning [21] which reflects the central position of the machine learning area of AI in industry interoperability requirements.

Of importance to the mapping of AI standards to industry practice and regulation was the approval in August 2020, after a justification study, to develop an AI Management System (AIMS) Standard [28]. Management System Standards have a distinct role on the ISO ecosystem of standards types, as they provide the basis for certifying organisational processes. This provides a basis for organisations to demonstrate their conformance to specific standardised behaviour for management and related technical operations processes. Regulatory authorities also can make reference to such standards in specifying compliance regimes in complex technical domains. This allows authorities to manage the complexity and risk of technological change in regulations and to do so in a way that aligns with international industry and society consensus established through international standards. In contrast with industry consortia active in standardisation, standards produced by ISO and IEC are driven by national bodies (ISO and IEC members) who are typically mandated by their governments to represent a wider range of societal stakeholders than just industry. The overarching goal of these member organisations is to ease doing business according to the United Nations World Trade Organisation's charter, as well as achieving United Nations' sustainable development goals (SDGs).

In recognition that big data plays a central role in the development of modern AI systems, Working Group 2 (WG 2) of SC 42 has developed a series of big data standards. This includes a Big Data Reference Architecture (BDRA) [29] that provides a structured set of functional areas related to Big Data processing. Currently, WG 2 is developing a process management framework for big data analytics [30].

Finally, SC 42 also hosts and leads a joint working group (JWG 1) with JTC 1/SC 40 which addresses IT service management and IT governance in a specification for governance implications of the use of artificial intelligence by organisations [31]. This builds on the existing SC 40 standard providing guidance and principles for the effective, efficient, and acceptable governance of IT in an organisation [32].

4. Conceptual modelling approach

The formalisation of terminology is already a key element of international standards development [33], Section 16. However, the use of ontologies to improve the consistency and coordination in international standards is not so well established, though it has been proposed [34]. There are some ambitious proposals to drive standards development together with implementation and verification from machine readable standards, with machine readable semantics of ontologies at its core [35], annex 11.4. The proposed application of ontologies to automated assistance in standards management and compliance, however, is outside the scope of this chapter.

ISO/IEC JTC 1/SC 32 has established requirements for ontologies to support semantic interoperability between information systems developed by different organisations and consortia across different application domains. These requirements address ontology definitions in a hierarchical manner with a top-level ontology that provides core concepts that can be used in defining more specialised domain-specific ontologies [36]. Such ontologies are expressed through a combination of natural language definitions and a machine-readable representation in a combination of description logic captured in the OWL2 language standard by the W3C [37] and Common Logic (CL) [38]. While CL is a full First Order Logic and therefore more expressive than OWL2, the latter offers the advantage of decidability by automated reasoners, therefore better supporting automated checking of specifications. For the top-level ontology, the Basic Formal Ontology (BFO) is used [39]. To support the mapping between different domain models, the BFO adopts a realist design philosophy that aims to capture the objective reality in a domain rather than existing data representations. The BFO places a strong emphasis on representing temporal and spatial characteristics of concepts. For example, at the highest level it distinguishes between conceptual hierarchies for continuant entities that persist over time, and occurrent entities that are time bound. Separate to its inclusion in ISO/IEC standardisation's use of ontologies, the BFO has been used for a set of biomedical ontologies [40] and for a collection of general common core ontologies addressing both horizontal concepts, known as mid-level ontologies, e.g., on event, information, currency, and domain-specific ontologies such as spacecraft and ethnicity [41].

While the BFO provides a precise conceptual structure for modelling a wide range of concepts, its realist approach to ontology engineering implies a need for a centrally controlled programme of ontology development to ensure consistent use of top-level concepts. While this would suit the objective of checking for and resolving consistency between a set of documents being developed under a common authority e.g. within SC 42, the requirements for checking consistency between regulatory and organisation policy drafts raises the needs to support a broad cohort of conceptual modellers who can operate without a common coordinating authority. An ontological approach that will be accessible for analysis and development by this wider range of conceptual modellers points therefore to demand less stringent conceptual modelling skills and is more accommodating to decentralised, domain-specific, and parallel development than BFO would enable.

The World Wide Web Consortium's (W3C) Data Activity (https://www.w3.org/2013/data/)buildson its earlier Semantic Web activity in developing industry standards for semantic interoperability that works to the scale and decentralised nature of the WWW. Grounded in the Resource Description Framework (RDF) [42] which allows an unlimited knowledge graph of nodes and links to exist as online resources on the WWW. Nodes and associations in this knowledge graph are typed according to ontologies, also known as data vocabularies, that can be developed independently and published to a distinct namespace on the WWW. This name-spaced typing allows the free combination of types and associated conceptual knowledge from any published vocabulary. This has enabled the development and integration of a global network of over 1200 open and interlinked data sets, known as the linked open data cloud [43]. Further logical axioms can be introduced to a vocabulary by including constructs from the Web Ontology Language (OWL) [37]. The W3C restricts its ontology standardisation activities to those vocabularies that support the development, location, interlinking and use of datasets on the web and to application areas that are well aligned to decentralised data publishing e.g. geospatial data. Organisations and consortia that develop vocabularies that address specific domains are free to publish them under their own name space, reusing aspects of other ontologies as needed. This highly decentralised approach therefore aligns well with the goal of promoting participation of those generating standards, organisational policies and regulations as well as those with an interest in how these documents develop and map to each other.

There is extensive research conducted on the use of ontologies in assessing compliance of organisational policies and practices with regulation [44]. Such assessment can be categorised as ex-ante compliance activities conducted before the regulated activity is performed or as ex-post compliance which is conducted after the regulated activities have occurred. The Provenance Ontology (PROV-O) is a W3C standard for the representation of provenance information using semantics provided by RDF and OWL [45]. PROV-O provides classes, properties, and restrictions to represent and interchange provenance information within different contexts. Its conceptual model is focussed on provenance of activities, the entities that those activities use and generate, and the actors associated with those entities or to whom activities are attributed. PROV-O can be specialised to create new classes and properties to model provenance information for different applications and domains. This has been utilised to extend its conceptual model to specify 'plans' for ex-ante representations, scientific workflows, and domain-specific requirements for activities. An example of its usefulness is the representation of ex-ante and ex-post activities for evaluating compliance with EU's General Data Protection Regulation [46], which is one of the more relevant extant regulations for AI. This existing capability, coupled with the decentralised conceptual modelling enables by the underlying RDF/OWL-based approach, leads us to adopt PROV-O, and its modelling of activities in particular, as the basis for conceptual modelling to support semantic interoperability between different sources of trustworthy AI concepts outline in Section 2.

5. Ontology for semantic interoperability between trustworthy AI documents

The approach therefore taken in this chapter combines elements of BFO and PROV-O to provide a minimal and easy to understand ontology for mapping between different standard, regulations and organisational policies. Specifically, we retain the distinction from BFO between continuant and occurrent entities. For the continuant concept we adopt the PROV-O concept of *Entity*, which is defined as: a physical, digital, conceptual, or other kind of thing with some fixed aspects; entities may be real or imaginary. For the occurrent concept we adopt the PROV-O concept of *Activity*, which is defined as: something that occurs over a period of time and acts upon or with *Entities*. It may include consuming, processing, transforming, modifying, relocating, using, or generating entities. We complement these core concepts of entity and activity with the PROV-O concept of *Actor* which is defined as: something that is involved in and may bear some form of responsibility for an *Activity* taking place, for the existence of an *Entity*, or for another *Agent's Activity*. BFO captures the organisational concept of a role as a subclass of continuant. However, the centrality of ethics and responsibility to the conceptual modelling of trustworthiness as well as for practical mapping to normative rules that are essential to standards, regulation and organisation policies demands that the concept of an *Agent* is made core alongside *Entity* and *Activity*. For the core relationships between these core concepts, we again draw from PROV-O relationships between these concepts, but recognising the benefit of the realist philosophy of BFO, these are phased in the present tense, rather than the past tense expression of PROV-O, which has a narrower focus on recording provenance i.e. what has existed and occurred. Therefore, we introduce the following associations:

- *Activity uses Entity*: where usage is defined in PROV-O as the beginning of utilisation of an *Entity* by an *Activity*, where before usage, the *Activity* had not begun to utilise this *Entity* and could not have been affected by the *Entity*.

- *Activity generates Entity*: where generation is defined by PROV-O as the completion of production of a new *Entity* by an *Activity*, where this *Entity* did not exist before generation and becomes available for usage after this generation. This therefore allows dependency chains to be assembled between *Activities*.

- *Entity attributedTo Agent*: which is the ascribing of an *Entity* to an *Agent*, thereby conveying some sense of responsibility for the *Entity* to that *Agent*. This provides a short form for relating an *Entity* to an *Agent* when the *Activity* associated with the *Actor* that generates or uses the *Entity* is not specified.

- *Activity associatedWith Agent*: which PROV defines as an involvement by an *Agent* in an *Activity*.

- *Activity informedBy Activity*: which allows a dependency between two *Activities* to be defined without needing to specify the generation and use of an *Entity* between them.

- *Agent actsOnBehalfOf Agent*: which is used in PROV-O to express delegation, which is the assignment of authority and responsibility to an *Agent* to act as delegate or representative of another *Agent*. The delegation can be assigned to a specific *Activity* which implies that the delegate *Agent* therefore takes on a degree of responsibility for the *Activity*.

- *Entity derivesFrom Entity*: where PROV-O defines a derivation as a transformation of one *Entity* into another, an update of an *Entity* resulting in a new one, or the construction of a new *Entity* based on a pre-existing *Entity*. This relationship can be associated with a specific *Activity*, offering a similar chaining relationship to *generates* and *uses*.

In addressing these core concepts and relationships, we express a scope related to the activities of an organisation that relate to the trustworthiness of AI. Trustworthiness can be manifested through *Trustworthy Characteristics* of *Entities* that make up a product or service employing AI. Such AI *trustworthiness characteristics* can be for example associated with: the datasets used to train data-driven machine learning AI [47–49]; trained ML models [50, 51] or AI-based product or services [52]. While some of these existing approaches share specific type of trustworthy characteristics for an *Entity*, e.g. those related to bias in ML models or the processing of personal data, these differ according to the needs of the specific link in the AI value chain that is being served, i.e. is the organisation providing dataset, AI model or completed AI-based products and service to its customers. However, the common feature of the above use of *Trustworthy Characteristics* is that they are communicated between one actor in a value chain and another. As the SC 42 technical report on an Overview of AI Trustworthiness [22] states that it is insufficient for one to simply refer to the 'trustworthy AI', but instead one must specify who trusts whom in what aspects of AI development and use. In this ontology, therefore, we define that the *Actor* that is providing an *Entity* to another *Actor* takes responsibility for ensuring that the *Trustworthy Characteristics* of the *Entity* in question is *exhibited*, and not necessarily just to the *Actor* receiving them but to any interested parties.

Trustworthiness can also be manifested through the *Activities* of an organisation involved in providing *Entities* that exhibit some *Trustworthy Characteristics*. The existing examples of approaches to expressing trustworthy characteristics of *Entities* referenced above define many of those characteristics in terms of *Activities* that were conducted to arrive at the characteristic, e.g. risk assessment.

To date, SC 42 has focussed on developing process-oriented standards, which is in line with the approach to developing Management Systems Standards (MSS) [53] annex SL. This approach is now being applied by SC 42 to develop a standard for an AI Management Systems (AIMS) [28]. A Management System is defined as a set of interrelated or interacting elements of an organisation to establish policies and objectives and processes to achieve those objectives. An MSS aims to provide organisations with guidance on standards activities in a manner that may be subject to independent certification. Such certification can also play an important role in establishing trust in an organisation's implementation of complex technical processes as part of a regulatory framework. Therefore, the following concepts from [28] are added to the model as subclasses of *Agent*:

- *Organisation*: person or group of people that has its own functions with responsibilities, authorities and relationships to achieve its objectives [28]. It is the *Organisation* which is the *Actor* disclosing *Trustworthy Characteristics*.

- *Stakeholder* (used synonymously in [28] with an 'interested party'): person or organisation that can affect, be affected by, or perceive itself to be affected by a decision or activity, that are conducted by the *Organisation* implementing an AIMS. The *Stakeholder* is therefore the *Actor* towards which any *Trustworthy Characteristic* is exhibited, with the aim of contributing to how that *Actor* determines its level of trust in what the *Organisation* claims possesses that characteristic.

Stakeholders can be internal or external to the *Organisation*. Therefore, an *Organisation* will be involved in disclosing *Trustworthy Characteristics* to both internal and external stakeholder. Internal stakeholders include those with overall organisational governance responsibilities, i.e., the governing body, management and employees. Internal stakeholder also includes those bound to an Organisation by contracts, which would include shareholders and supply chain partners, i.e., providers, customer and individual consumers. External Stakeholders can include regulators, government, potential supply chain partners, consumers, civic society, NGOs, the media and society in general.

As *Trustworthy Characteristics* are defined as a relationship between *Organisations* and *Stakeholders*, *Entities* being characterised are subclassed *Assets*. *Assets* are *Entities* that represent some value to a *Stakeholder*. For AI internal stakeholders, Assets could include the data used to train or test an AI system, a trained AI model, a product or service that uses one or more AI systems, data consumed by an AI system in operation, software, computing and human resources used in training and operating an AI system [22].

In considering the trustworthiness of an AIMS we would be focussed on the *Activities* conducted by the *Organisation* operating the AIMS. However, this subclassing of *Agents* also allows *Activities* with *Trustworthy Characteristics* to be associated with *Stakeholder*, and in particular external stakeholders, which allows complementary activities, e.g., in value chain partners or regulators, to be modelled. While external Activities may not be an explicit part of an AIMS specification, they may be referenced in regulations or organisational policies which aim to clarify the interactions and share of responsibilities between *Activities* of the *Organisation* and those of such external stakeholders.

To capture to different responsibilities within an Organisation operating an AIMS, additional semantics for the relationships between Activities (beyond dependencies via the use and generation of Entities and the *dependsOn* relationship) are required. Reference architectures and process model specifications often group activities into groups representing closely related or commonly co-occurrent process. Activities are also sometime grouped into roles, which are a set of activities servicing some common purpose, the responsibility for which is often allocated together to a specific organisational unit. We therefore include a compositional relationship between *Activities* called *partOf* that allows for various forms of Activity groupings.

Within the context of an AIMS, and its mapping to regulations and policies of specific organisations, it is also important to capture some sense of different levels of responsibility and corresponding accountability. SC40 distinguishes between an organisation's governance function and its management function as part of guidelines the governance of IT [32]. The governance function is responsible for understanding the external pressures in forming an organisation's direction, strategy, objectives and policy in adopting IT, including customer, competitive, stakeholder expectation and regulatory pressures. The management function is then responsible for planning and achieving those objectives within the constraints of the strategy and policy. The governance function is structured into three activities, which are elaborated in the governance of IT implementation guide standards [54], Evaluate, Direct and Monitor. The evaluate activity addresses internal and external considerations in assessing plans and proposals from the management function. The direct activity sets direction through strategy and policy and assigns responsibilities for their realisation by the management function. The monitor activity assesses the performance of the realisation of governance direction on a regular basis, triggering further evaluation and direction if deficiencies are identified. Management functions are responsible for the control of technical operations activities and may do so through delegation between appropriate levels of control. To capture the relationship between governance and management activities and between different levels of control between management activities, the following relationships between Activities in the ontology are defined: *directedBy*, *evaluatedBy*, *monitoredBy* and *controlledBy*.

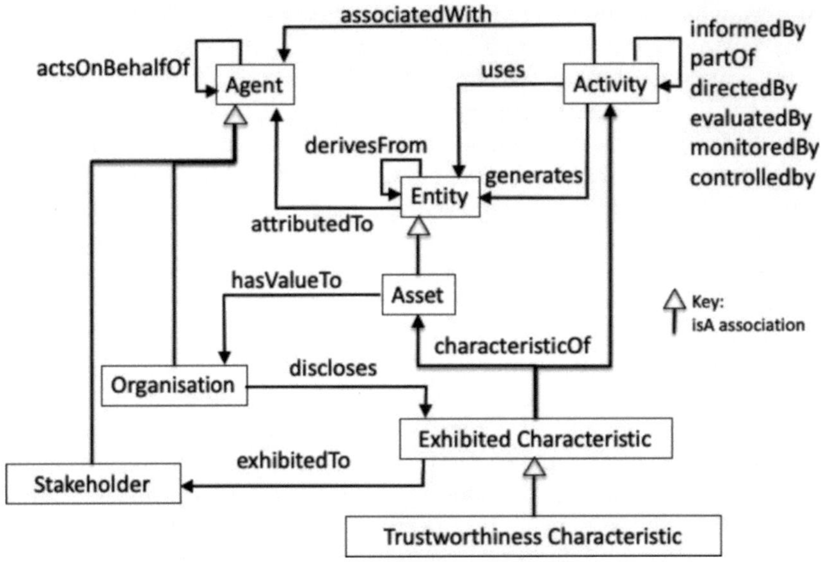

Figure 2.
Core concepts and relationship for semantic interoperability for trustworthy AI.

Figure 2 captures the above concepts and relationships into a core ontology that is intended to support the mapping of concepts between different emerging AI standards in SC 42 and between those standards and emerging organisational policies or governmental regulations as indicated in **Figure 1**. By restricting these concepts to a small core that is already established in existing standards, upon which some of the SC 42 standards are based, we anticipate this ontology will provide a robust basis for identifying the relationships between such concepts in different specifications.

SC 42 has already identified characteristics that a trustworthy AI system or organisation involved in their implementation could exhibit [22]. These include technical characteristics such as reliability, robustness, verifiability, availability, resilience, quality, bias and robustness; stakeholder-related characteristics such as ethics, fairness and privacy; as well as management- and governance-related characteristics such as transparency, explainability, accountability and certification. However, the definition of many of these characteristics are still not yet well defined in relation to AI. By focussing on Activities, Entities and Agents we aim to identify mappings between concepts in different standards and therefore any gaps that can directly inform more consistent and comprehensive standards. In this way we hope to assist in the progression from broad statements of principles and areas of concern by the private sector, international bodies and governments, towards the develop of commonly understood process framework for the governance and management of Trustworthy AI, the ontology aims to do this is a way that accommodates the current range of definitions and interpretation of many of these characteristics and supports their convergence over time into concrete and internationally recognised governance and management processes and policies. In the following sections we show how this process can be undertaken by using this ontology to map between activities in the core anticipated AIMs and other relevant SC 42 specifications.

6. Modelling of AI management system activities

In 2020, SC 42 completed a justification study for a Management System Standard for AI. This was accepted in August and led to the initiation of a project

Activity ID	label	See also "42001-*"	attributedTo	generates (type Entity)	uses (type Entity, "aims*")
aimsA1	Understanding organisation and its concepts	4.1	O	aimsE1 - External and internal issues affecting AIMS outcomes	
aimsA2	Understanding stakeholder needs and expectation	4.2	O	aimsE2 - AIMS Stakeholders; aimsE3 - Stakeholder requirements	E1
aimsA3	Determine AIMS scope	4.3	O	aimsE4 - AIMS Scope	E1, E2, E3
aimsA4	Establish AIMS policy	5.2	TM	aimsE5- AI Policy	E1, E2, E3
aimsA5	Assign roles, responsibilities and authorities	5.3	TM	aimsE6- Roles, responsibilities & authority assignments	E3, E4, E5
aimsA6	Address risks and opportunities	6.1	O	aimsE7- AI risk management plan	E1, E2, E3
aimsA7	Establish and plan to achieve AI objectives	6.2	O	aimsE8- AI objectives; aimsE9- Plan to achieve AI objectives	E3, E4, E5, E6
aimsA8	Determine and allocate resources for AIMS	7.1	O	aimsE10- AI resource allocation	E9
aimsA9	Determine and ensure competence of people affecting AI performance	7.2	O	aimsE11- AI competence plan	E9
aimsA10	Promote awareness	7.3	O	aimsE12 - AI awareness plan	E3, E5, E6, E9
aimsA11	Determine AIMS communication	7.4	O	aimsE13 - AI communication plan	E3, E9
aimsA12	Plan and control AI processes	8.1	O	aimsE14 - AI operational process control plan	E6-E9
aimsA13	Monitor, measure, analyse and evaluate AI	9.1	O	aimsE15 -AI & AIMS evaluation plan; aimsE16 - AI & AIMS evaluation results	E3-E8
aimsA14	Internal audit	9.2	O	aimsE17 - AIMS audit plan; aimsE18 - AIMS audit results	E3-E6, E15, E16

Activity ID	label	See also "42001-*"	attributedTo	generates (type Entity)	uses (type Entity, "aims*")
aimsA15	Undertake management review	9.3	TM	aimsE19 - AIMS change and improvement report	E1-E19
aimsA16	Detect non conformance and take corrective action	10.1	O	aimsE20 - AIMS non conformity and corrective action report; aimsE21 - AIMS corrective action evaluation	E5, E8, E7
aimsA17	AIMS Continual improvement	10.2	O	aimsE1 - aimsE15; aimsE17; aimsE20; aimsE21	E1-E15

** = wildcard representing row value(s) in the column.*

Table 1.
Activity and entities identified for an AIMS.

to develop as AI Management System (AIMS) Standard [28], following the guidelines set out in the ISO/IEC Directive 1 [53]. All MSS should follow a consistent high-level structure which includes common text and terminology as presented in Annex SL of this Directive. This is to allow different MSS addressing different horizontal and domain-specific areas to be integrated into the same overall management system within an organisation that implements these MSS. It is intended that MSS should be developed in a manner open to different stakeholders, including accreditation bodies, certification bodies, enterprises and the user community. The high-level structure for MSS therefore provides a well-defined source of concepts that are likely to be reflected in the MSS being developed for AI, which must also address management aspects of trustworthiness. The use of MSS for organisational process certification means it is also a suitable source of concepts that will be useful to track against those being developed for public and organisational policy and processes.

Table 1 presents a mapping of the MSS high level structure as a set of 17 AIMS activities, with each concept given an identifier, a label attribute, and a 'see also' attribute referencing the relevant section in the AI MSS draft, the numbering for which is mandated in the MSS high level structure. Each of these AIMS activities is attributed to either the organisation overall (O) or top management (TM) as defined in the MSS high level structure, where the former attribution implies that activity spans governance and management levels. Relationship between these activities is captured by generates and uses attributes referencing 21 AIMS entities derived from the text in the MSS high level structure.

7. Modelling of AIMs technical operation activities

The foundational standard on AI terms on concepts [20] entered its second committee draft ballot review towards the end of 2020 and is therefore still under development. However, Section 6 of the draft does provide an outline of the life-cycle for an AI system made up of activities: inception; design and development,

ID 22989*	CD2 22989 - AI terms and concepts	partOf
IC v	Inception	
IC1	Determine stakeholders' objectives	IC; aimsA1
IC2	Determine stakeholders' requirements	IC; aimsA2
IC3	Risk assessment and treatment planning	IC; aimsA6
IC4	Policies and compliance planning	IC; aimsA4
DD ^*	Design and development	
VV ^*	Verification and validation	
DE ^*	Deployment	
OM v*	Operation and monitoring	
OM1	Monitor AI system	OM; aimsA13
OM2	Repair AI system	OM
OM3	Update AI system	OM
OM4	Support AI system users	OM; aimsA11
OM5	OM5: Risk monitoring and review	OM; aimsA6
CV ^*	CV: Continuous validation	
RE ^*	RE: Re-evaluation	
RT ^*	RT:: Retirement	

Table 2.
*Partially expanded view of AI lifecycle activities and sub-activities from CD22989 [20]: Activity ID key: v = constituent activities expanded, ^ = constituent activities collapsed, * = AI technological operation activities.*

verification and validation; deployment; operation and monitoring; continuous validation; re-evaluation; and retirement. **Table 2** shows a partially expanded breakdown of sub-activities that are part of these activities, exemplifying how the partOf attribute of these activities are used to identify the constituent activities. It also identifies certain sub-activities as also part of specific AIMS activities from **Table 1**. Sub-activities under the Inception activity are identified as part of specific AIMS activities which would form part of the governance and management activities. The operations and monitoring activity (OM in **Table 2**) of the AI lifecycle also contains some sub-activities that are part of AIMS activities, related to monitoring (aimsA13), communication (aimsA13) and risk (aimsA6). Further, all the AI lifecycle activities except Inception, are classified as technical operations activities, meaning within the AIMS they are directed by activity aimsA3 and aimsA4, evaluated by activities aimsA8 and aimsA9, monitored by activities aimsA13, aimsA14, aimsA15, aimsA16 and aimsA17 and controlled by activities aimsA12 (plan and control AI processes). The transitive nature of the partOf relationship therefore implies that the sub-activities of these AI lifecycle activities are also classed as technical operation activities with the same directedBy/evaluatedBy/monitoredBy/controlledBy relationship to the corresponding AIMS activities. As AIMS activities can operate at different levels of management delegation, technical operations activities such as operations and monitoring (OM) therefore both play a part in conducting AIMS activities, but at a much narrower level of abstraction, as well as being subject to direction, evaluation and monitoring of other AIMS activities. It is notable that all the top layer activities in this lifecycle model, apart from retirement activities (RT) are part of the AIMS activity address risk (aimsA6).

Though this lifecycle model from [20] is still a draft, it is being used in other SC 42 drafting activities, specifically the technical report on AI bias [23], which in Section 9 breaks down bias treatment at each of the top levels activities of the lifecycle model, and a new work item on AI system life cycle processes to support their definition, control and improvement within an organisation or project [55].

As well supporting the mapping of AIMS activities into standards that explicitly define processes that we can model as activities, the ontology from **Figure 2** can also be used to map AIMS into role-based frameworks. Specifically, SC 42 Big Data Reference Architecture [29], which is structured into big data provider roles of: application provider, framework provider, service provider, data provider and consumer. These roles further are subdivided into sub-roles for which constituent activities are defined. This conforms with the notion of a role being a set of activities, so again this containment structure can be expressed as partOf relationship between the different levels of activity sets corresponds to those roles. All these are categories as technical operations activities in relation to the AIMS activities, though some also map to specific AIMs activities, with roles-based activities related to auditing (aimsA14) and requirements capture (aimsA7).

This mapping of activities reveals that while different standards developed under SC 42 are broadly consistent with the high-level structure of AIMS, they vary in which areas of AIMS they focus on. While this may be arguably appropriate for the projects concerned, it does indicate that AIMS activities are not comprehensively mapping into other SC 42 standards as a matter of course. The ontology therefore provides a way of tracking these mapping and identifying gaps that may need to be addressed as the AIMS is specified or as changes to the other standards as they develop or are reviewed over time.

8. Modelling stakeholders

The mapping of role-oriented activity processes from the Big Data Reference Architecture to AIMS activities also indicates how groupings of activities expressed as a role, e.g. Big Data Application Provider or Big Data Consumer, can also be used to model a stakeholder. The draft foundational terms and concepts standard from SC 42 [20] has similarly identified stakeholder in relation to similar types of value-chain role, but SC 42 has not yet attempted a more detailed characterisation of the activities that such stakeholders would undertake, and which would therefore define such stakeholder roles.

However, many aspects of AI ethics and their impact on the trustworthiness of AI, relate to the impact on stakeholders who are not directly involved in the AI value chain, even as customers or consumers, e.g., pedestrians injured by an automated vehicle, local communities blighted by automated routing of commercial traffic through their neighbourhood, or people denied access to financial, health or social security services due to bias in algorithmic decision-making. In addition, such indirectly affected stakeholders may be difficult for organisations to identify and consult, so appropriate representation may be needed. Such representation could take the form of NGOs concerns with rights of certain groups, labour unions, professional bodies, local community groups, up to and including democratic forms of representation in the form of local and national governments. These issues are addressed however in the ISO Guidelines for Social Responsibility ISO26000 [56]. Its guidance is based on the fundamental practices of recognising social responsibility within an organisation and undertaking stakeholder identification and engagement. The guidance is based on the principles similar to those identified in the growing literature on AI ethics [6], including accountability, transparency and ethical behaviour as well as respect for the rule of law, international norms of behaviour and human rights. Social responsibility guidance is provided in terms of 37 issues, each with suggested actions to address them and associated behavioural exceptions. These issues are each grouped under one of the following social responsibility core subjects: Human Rights; Labour practices; the Environment; Fair operating practices; Consumer issues; and Community involvement and development. The fact that these core subjects map onto different areas of legislation in many jurisdictions internally also eases the mapping of guidelines using this structure onto regulatory or other legal obligations that must be complied with by an organisation's management system. ISO26000 does not specifically address the use of AI, however these issues are sufficiently broad to provide a basis for mapping out specific ethical and societal issues associated with AI as shown through a comparison [57] with the normative language on trustworthy AI principles established in [2, 4, 5]. ISO already provides guidance in [58] on how to map principles and issues from ISO 26000 to the high level structure of MSS. SC 42 also recognises the importance of ISO 26000 guidance in handling stakeholder issues in the draft standards on AI governance [31] and AI risk management [24], with the draft technical report on ethical and societal concerns of AI [22] including a high level mapping of ISO26000 core issues onto AI risks and treatments.

9. Conclusions and further work

This chapter has highlighted the multiplicity of parallel activities being undertaken in developing international standards, regulations and individual organisational policies related to AI and its trustworthiness characteristics. The current lack of mappings between these activities presents the danger of a highly fragmented global landscape emerging in AI trustworthiness. This could present society, government and industry with competing standards, regulations and organisational practices that will then serve to undermine rather than build trust in AI. This chapter presents an overview of AI standardisation currently being undertaken in ISO/IEC JTC 1/SC 42 and identifies its work to define an AI management system standard as the starting point for establishing conceptual mapping between different initiatives. A minimal, high level ontology for the support of conceptual mapping between different standardisation, regulatory and organisational policy documents is presented. We show how this can help map out the overlaps and gaps between AI governance, management and technical operations activities present in some of the SC 42 standards currently under development.

Further work is required to develop and maintain a mapping between the ontological concepts and relationships identified from the emerging set of SC 42 AI standards and the emerging trustworthy AI regulations and policies from different organisations. The mapping of such standards to the ontology could be made publicly available in a findable, accessible, interoperable and reusable form, using linked open data principles [43], and updated as the referenced specifications evolve. This will assist in identifying gaps and inconsistencies between evolving drafts, especially in developing the AIMS standard [20]. The set of trustworthy AI characteristics could be captured in the ontology, based in the first instance on the AI engineering quality characteristic being developed in [26]. Similarly, the ontology can be extended to express sets of AI risks and treatments so concepts developed in AI risk [24] and bias [23] will also be captured.

The use of this ontology-based approach for comparing the guidance between standards could also be applied between SC 42 and the largely orthogonal set of standards being developed under P7000. These include ethical design processes, transparency for autonomous systems, algorithmic bias, children, student and employee data governance, AI impact on human well-being, and trustworthiness rating for news sources.

Draft legislations for AI such as [59] will need to be analysed in terms of activities, actors, entities, characteristics and risks so that a mapping to the equivalent concepts from the SC 42 specifications family can be assembled and maintained. Similar analyses will be undertaken on publicly available policies from international bodies such as the EU High Level Expert Group on AI's checklist for trustworthy AI [60] and the proposals emerging from the private sector for assigning trustworthiness declaration to products and services [47–51].

Acknowledgements

This work was conducted by the ADAPT Centre with support of SFI, by the European Union's Horizon 2020 programme under the Marie Skłodowska-Curie Gran Grant Agreement No. 813497 and by the Irish Research Council Government of Ireland Postdoctoral Fellowship Grant GOIPD/2020/790. The ADAPT SFI Centre for Digital Content Technology is funded by Science Foundation Ireland through the SFI Research Centres Programme and is co-funded under the European Regional Development Fund (ERDF) through Grant # 13/RC/2106_P2.

Conflict of interest

David Filip is the convener of ISO/IEC JTC 1/SC 42/WG3 and Dave Lewis is a contributor to several of the SC 42 standardisation projects covered in this chapter.

Thanks

Thanks to Delaram Golpayegani, Arthit Suriyawongkul and PJ Wall of the ADAPT Centre for their pertinent discussion and comments on the subject of this chapter.

Author details

Dave Lewis[1]*, David Filip[2] and Harshvardhan J. Pandit[1]

1 ADAPT Centre, Trinity College Dublin, Ireland

2 Huawei Technologies, Ireland

*Address all correspondence to: dave.lewis@adaptcentre.ie

References

[1] Adamson, G., Havens, J. C., & Chatila, R. "Designing a Value-Driven Future for Ethical Autonomous and Intelligent Systems", Proceedings of the IEEE, 107(3), 518-525. 2019, DOI: 10.1109/JPROC.2018.2884923

[2] "Ethically Aligned Design – First Edition", The IEEE Global Initiative on Ethics of Autonomous and Intelligent Systems, IEEE, 2019. available at: https://standards.ieee.org/industry-connections/ec/ead-v1.html

[3] ISO GUIDE 82:2019 Guidelines for addressing sustainability in standards, Status available at: https://www.iso.org/standard/76561.html

[4] HLEG (2019) European Commission's High Level Expert Group, "Ethics Guidelines for Trustworthy AI", April 2019, https://ec.europa.eu/futurium/en/ai-alliance-consultation/guidelines

[5] OECD (2019) OECD/LEGAL/0449 "Recommendation of the Council on Artificial Intelligence", https://legalinstruments.oecd.org/en/instruments/OECD-LEGAL-0449

[6] Jobin, A., Ienca, M. & Vayena, E. The global landscape of AI ethics guidelines. Nat Mach Intell 1, 389-399 (2019). DOI: 10.1038/s42256-019-0088-2

[7] Mapping regulatory proposals for artificial intelligence in Europe. Access Now, 2018 https://www.accessnow.org/mapping-regulatory-proposals-AI-in-EU

[8] Wagner, B., Ethics as an Escape from Regulation: From ethics-washing to ethics-shopping. In BEING PROFILED: COGITAS ERGO SUM: 10 Years of Profiling the European Citizen, edited by EMRE BAYAMLIOĞLU et al., Amsterdam University Press, Amsterdam, 2018, pp. 84-89. JSTOR, www.jstor.org/stable/j.ctvhrd092.18

[9] Calo, R. (2017). Artificial Intelligence Policy: A Roadmap. SSRN Electronic Journal, 1-28. DOI: 10.2139/ssrn.3015350

[10] Whittlestone, J., Nyrup, R., Alexandrova, A., Cave, S., The Role and Limits of Principles in AI Ethics: Towards a focus on tensions, in proc Artificial Intelligence, Ethics and Society, Honolulu, Hawaii, USA, Jan 2019

[11] Hagendorff, T. The Ethics of AI Ethics: An Evaluation of Guidelines. Minds & Machines 30, 99-120 (2020). DOI: 10.1007/s11023-020-09517-8

[12] Latzer, M., Hollnbuchner, K., Just, N., & Saurwein, F. (2016). The economics of algorithmic selection on the Internet. Handbook on the Economics of the Internet, (October 2014), pp 395-425. Retrieved from https://doi.org/10.4337/9780857939852.00028

[13] Tutt, A. (2016). An FDA for Algorithms. SSRN Electronic Journal, 83-123. https://doi.org/10.2139/ssrn.2747994

[14] Erdelyi, O.J. and Goldsmith, J., Regulating Artificial Intelligence: Proposal for a Global Solution (February 2, 2018). 2018 AAAI/ACM Conference on AI, Ethics, and Society (AIES '18), February 2--3, 2018, New Orleans, LA, USA. DOI: 10.1145/3278721.3278731, Available at SSRN: https://ssrn.com/abstract=3263992

[15] Calo, R. (2013). Consumer Subject Review Boards: A Thought Experiment. Stanford Law Review Online, 66(2002), 97.

[16] Polonetsky, J., Tene, O., & Jerome, J. (2015). Beyond the common rule: Ethical structures for data research in non-academic settings. Colo Tech L J, 13, 333-368.

[17] Scherer, M.U., (2016) Regulating Artificial Intelligence Systems: Risks,

Challenges, Competencies and Strategies, Harvard Journal of Law & Technology, v29, no 2, Spring 2016

[18] Schiff, D., Borenstein, J., Biddle, J., Laas, K., "AI Ethics in the Public, Private, and NGO Sectors: A Review of a Global Document Collection," in IEEE Transactions on Technology and Society, doi: 10.1109/TTS.2021.3052127

[19] ISO/IEC TR 24028:2020 Information technology — Artificial intelligence — Overview of trustworthiness in artificial intelligence. 2020. Available from: https://www.iso.org/standard/77608.html

[20] ISO/IEC CD 22989.2, Artificial intelligence — Concepts and terminology, status available at: https://www.iso.org/standard/74296.html

[21] ISO/IEC CD 23053.2 Framework for Artificial Intelligence (AI) Systems Using Machine Learning (ML). Status available at: https://www.iso.org/standard/74438.html

[22] ISO/IEC AWI TR 24368 Information technology — Artificial intelligence — Overview of ethical and societal concerns. Status available at https://www.iso.org/standard/78507.html

[23] ISO/IEC DTR 24027 Information technology — Artificial Intelligence (AI) — Bias in AI systems and AI aided decision making. Status available at: https://www.iso.org/standard/77607.html

[24] ISO/IEC CD 23894 Information Technology — Artificial Intelligence — Risk Management. Status available at: https://www.iso.org/standard/77304.html

[25] ISO 31000:2018 Risk management — Guidelines. Available at: https://www.iso.org/standard/65694.html

[26] ISO/IEC AWI 25059 Software engineering — Systems and software Quality Requirements and Evaluation (SQuaRE) — Quality model for AI-based systems. Status available at: https://www.iso.org/standard/80655.html

[27] ISO/IEC AWI 24029-2 Artificial intelligence (AI) — Assessment of the robustness of neural networks — Part 2: Methodology for the use of formal methods. Status available at: https://www.iso.org/standard/79804.html

[28] ISO/IEC AWI 42001 Information Technology — Artificial intelligence — Management system. Status available at: https://www.iso.org/standard/81230.html

[29] ISO/IEC 20547-3:2020 Information technology — Big data reference architecture — Part 3: Reference architecture. Available from: https://www.iso.org/standard/71277.html

[30] ISO/IEC WD 24668:2020 Information Technology — Artificial intelligence — Process management framework for big data analytics. status available at: https://www.iso.org/standard/78368.html

[31] ISO/IEC CD 38507 Information technology — Governance of IT — Governance implications of the use of artificial intelligence by organizations. Status available at: https://www.iso.org/standard/56641.html

[32] ISO/IEC 38500:2015 Information technology — Governance of IT for the organization. Available at: https://www.iso.org/standard/62816.html

[33] ISO/IEC Directives, Part 2 Principles and rules for the structure and drafting of ISO and IEC documents, 8th Edition 2018, Available from: https://www.iso.org/directives-and-policies.html

[34] Narayanan, A., Lechevalier, D., Morris, K.C., Rachuri, S. Communicating standards through structured

terminology, Computer Standards & Interfaces, Vol 40, 2015, Pages 34-41, ISSN 0920-5489, DOI: 10.1016/j.csi.2015.02.004

[35] Wahlster, W., Winterhalter,C., German Standardisation Roadmap for Artificial Intelligence, DIN & DKE, Nov 2020, Available from: https://www.din.de/resource/blob/772610/e96c34dd6b12900ea75b460538805349/normungsroadmap-en-data.pdf

[36] ISO/IEC DIS 21838-1:2019 Information Technology - Top-level ontologies (TLO) - Part 1: Requirements

[37] World Wide Web Consortium W3C Recommendation - OWL 2 Web Ontology Language Language Overview (2nd edition) available at: http://www.w3.org/TR/2012/REC-owl2-overview-20121211/

[38] ISO/IEC 24707:2018, Information Technology - Common Logic (CL): A framework for a family of logic-based languages, status available at: https://www.iso.org/standard/66249.html

[39] ISO/IEC: DIS 21838-2:2019 Information Technology - Top-level ontologies (TLO) - Part 2: Basic Formal Ontology (BFO)

[40] Smith B, Ashburner M, Rosse C, et al. The OBO Foundry: coordinated evolution of ontologies to support biomedical data integration. Nat Biotechnol. 2007; 25(11):1251-1255. DOI:10.1038/nbt1346

[41] Rudnicki, R, An Overview of the Common Core Ontologies, CUBRC, Inc. Feb 2019, available from: https://www.nist.gov/system/files/documents/2019/05/30/nist-ai-rfi-cubrc_inc_004.pdf

[42] RDF 1.1 Primer,. Available: https://www.w3.org/TR/rdf11-primer/ (visited on 08/11/2019).

[43] Bizer, C., Heath, T., Berners-Lee,T., "Linked Data: The Story so Far", Semantic Services, Interoperability and Web Applications: Emerging Concepts, pp. 205-227, 2011. DOI: 10/dh8v52. [Online]. Available: https://www.igi-global.com/chapter/linkeddata-story-far/55046 (visited on 09/17/2019).

[44] Fellmann, M., Zasada, A. State-of-the-Art of Business Process Compliance Approaches: A Survey, Proceedings of the 22nd European Conference on Information Systems (ECIS), Tel Aviv, June 2014

[45] Lebo, T., Sahoo, S., McGuinness, D., Belhajjame, K., Cheney, J., Corsar, D., Garijo, D., Soiland-Reyes, S., Zednik, S., Zhao, J., (2013). PROV-O: The PROV Ontology, http://www.w3.org/TR/2013/REC-prov-o-20130430/

[46] Pandit, H.J., . O'Sullivan, D., and D. Lewis, D., "Queryable Provenance Metadata For GDPR Compliance", in Procedia Computer Science, ser. Proceedings of the 14th International Conference on Semantic Systems 10th – 13th of September 2018 Vienna, Austria, vol. 137, Jan. 1, 2018, pp. 262-268. DOI: 10/gfdc6r. [Online]. Available: http:/ /www.sciencedirect.com/science/article/pii/S1877050918316314 (visited on 10/18/2018).

[47] Gebru, T., Morgenstern, J., Vecchione, B., Vaughan, J.W., Wallach, H., Daumé, H., Crawford, K. "Data sheets for Datasets, in Proceedings of the 5th Workshop on Fairness, Accountability, and Transparency in Machine Learning", Stockholm, Sweden, 2018.

[48] Bender, E., Freidman, B., "Data Statements for Natural Language Processing: Toward Mitigating System Bias and Enabling Better Science", in Transactions of the Association for Computational Linguistics, vol. 6, pp. 587-604, 2018

[49] Holland, S., Hosny, A., Newman, S., Joseph, J., Kasia Chmielinski, K., (2018) The Dataset Nutrition Label: A Framework To Drive Higher Data Quality Standards. arXiv:1805.03677

[50] Arnold, M., et al., "FactSheets: Increasing trust in AI services through supplier's declarations of conformity," in IBM Journal of Research and Development, vol. 63, no. 4/5, pp. 6:1-6:13, 1 July-Sept. 2019, doi: 10.1147/JRD.2019.2942288.

[51] Mitchell, M., Simone Wu, S., Zaldivar, A., Barnes, P., Vasserman, L., Hutchinson, B., Spitzer, E., Raji, I.D., Gebru, T.,. 2019. Model Cards for Model Reporting. In Proceedings of the Conference on Fairness, Accountability, and Transparency (FAT* '19). Association for Computing Machinery, New York, NY, USA, 220-229. DOI:https://doi.org/10.1145/3287560.3287596

[52] Hallensleben et al, From Principles to Practice: An interdisciplinary framework to operationalise AI ethics, AI Ethics Impact Group, April 2020, Available from: https://www.ai-ethics-impact.org/en

[53] ISO/IEC Directives, Part 1, Consolidated ISO Supplement — Procedures specific to ISO, 11th edition 2020. Available from: https://www.iso.org/directives-and-policies.html

[54] ISO/IEC TS 38501:2015 Information technology — Governance of IT — Implementation guide, Available from: https://www.iso.org/standard/45263.html

[55] ISO/IEC WD 5338 Information technology — Artificial intelligence — AI system life cycle processes, Available from: https://www.iso.org/standard/81118.html

[56] ISO 26000:2010 Guidance on social responsibility, status available at: https://www.iso.org/standard/42546.html

[57] Lewis, D., Hogan, L., Filip, D., P. J. Wall, PJ., Global Challenges in the Standardization of Ethics for Trustworthy AI, Journal of ICT Standardization, vol 8, iss2, 2020, DOI: 10.13052/jicts2245-800X.823

[58] IWA 26:2017 Using ISO 26000:2010 in management systems, status available at: https://www.iso.org/standard/72669.html

[59] A9-0186/2020 REPORT with recommendations to the Commission on a framework of ethical aspects of artificial intelligence, robotics and related technologies (2020/2012(INL)), Committee on Legal Affairs, European Parliament, 8.10.2020.

[60] HLEG European Commission's High Level Expert Group, "The Assessment List for Trustworthy Artificial Intelligence (ALTAI) for self assessment", July 2020.

Chapter 6

How Factoring Ethics Encourages and Stimulates Innovative Development of IT Systems Responsive to Stakeholder Needs and Requirements

Zvikomborero Murahwi

Abstract

Human beings have become increasingly dependent on IT in running their daily lives and doing business. The development and the increase in use of Autonomous Intelligent Systems in the last few years is making it increasingly impossible to ignore ethics in engineering and building IT Systems as more factors other than the traditional ones like financials come into play. As the use of technology continues to grow among people of all ages, there is also a growing awareness of the potential social harms such systems can have on human well being. Where there is such an awareness (of potential harms the systems may have on humans), there is a likelihood of resistance in adoption and use of the technology resulting in the loss of the benefits such systems are supposed to bring with them. Where potential harms have been ignored and people went ahead and adopted and used the technology, but eventually experienced cases of social harm, possible abandonment of the technology becomes a reality - a situation which also results in the loss of all the gains the technology could have potentially brought. Some good examples are societies which continue to keep under-aged children away from technology as a way of safeguarding them from the harms caused by technology. Situations like this have set limits to the effectiveness and reach of technology driven services such as autonomous eHealth systems and even educational programmes.There has also been a view mainly among technology creators that Ethics considerations are slowing down or getting into the way of innovation. In the proposed chapter, and driven by the theme "Ethics Considerations to drive innovative thinking in building systems which are responsive to the needs of the user and promote adoption, and safe use of IT" the writer argues for a unified front in the development of IT systems in an ethical manner - That is, it is time users and creators of technology start working together to build systems for the future. And with well-being among the UN Development Goals, and technology taking centre stage in many aspects of development and growth, Ethics in technology cannot be ignored.

Keywords: ethics, IT Systems, values, well-being

1. Introduction

Information Technology and specifically AI holds the potential to be a major driver of economic growth and social progress, if industry, civil society, government, and the public work together to support development of the technology with thoughtful attention to its potential and to managing its risks.

On 27 December 2020, Elon Musk tweeted 'Focus on making products and services people truly love, so that the total economic pie is bigger, instead of engaging in zero-sum / negative-sum corporate power struggles. This is the good future'. Of the 200 responses to this tweet read, only 3 did not directly agree. Some notable replies and comments to this tweet were:

- Profit Maximisation = Better products/services = Customer Retention + New Customers

- His logical approach of being a consumer, his passion, and futuristic vision for humanity speaks volumes for his products and companies. Corporations ran by MBA's are profit driven and have no belief in their product. They are not the same

- This is the root cause of why many innovative organisations atrophy over time. After success, they attract people who are more focused on the pay off than making something great. A team that truly cares about the product is often the number 1 competitive advantage in the long run

- After all, how is value created

- I suspect most companies do not have a strong, clear, or inspirational enough mission statement, otherwise this would be less of a problem

- What Elon says here is perfect. This should be read out at every board/senior staff/ offsite meeting that companies to set strategy. Unfortunately, legacy companies in industries getting disrupted do not think like this and that's why they are in so much trouble

- Well said, build great products that build value to people

- I agree but you need to really think all your products, process and people are the best innovations ever. Many great process and product innovations has not been hired for you

- This right here is why major companies used to live 60+ years. They made the consumer the general focus. Now its all about doing things the cheapest way so that all the corporate Titans can deepen their pockets which has nosedived the life expectancy of major companies to 18 years.

In response to a recent proposal to deploy an Electronic Health (e-Health) System for use with young people aged between 15 and 25, an East African NGO requested that alterations be made to the age group include to only those aged 18 and above because the country's laws did not allow those under the age of 18 to own personal mobile devices which were to be the main platform for deployment of the proposed solution. This meant that although studies had concluded that some under-18 age groups needed interventions provided for by the autonomous e-health system, those groups would be left out in deployment of the solution because of regulatory and cultural reasons.

Elon's tweet and replies to, and the above e-Health system scenario raise the following important issues in delivery of IT Systems and Services which are the subjects of discussions in this chapter:

- Understanding Values and Ethical Considerations in Systems, Products and Services
- Why values should be seriously considered when building IT systems and products
- How to incorporate values into IT systems and products and how this promotes innovation
- Implementing values and ethics in an organisation is a collective endeavour requiring active participation from all stakeholders
- Why and how incorporating values into systems and products promotes innovation

2. Concepts: values and ethics and why they matter

2.1 Understanding values

This chapters uses the Organisational Behavioural definition which defines values as the collective conceptions of what is considered good, desirable and proper or bad, undesirable, and improper in a culture or setting. That values set the standard for what one subscribes to or chooses. For example some common business values would be fairness, innovations and community involvement. It is not always easy to clarify the fundamental values of a given grouping, setting or society because of sheer breadth.

2.1.1 Some characteristics of values

- They can be different for each person or cultural setting.
- The degrees to which they are valued differs
- They inspire and motivate
- Normally stable but can change e.g. can change over time
- Can be specific to a person or situation (e.g.) or general (e.g. health, love)
- Core values can be learnt from family, neighbourhood and various sources within a setting or society
- Where common values are shared, they build up societies and integrate social relations

- They influence people's behaviour providing standards for performance and morality, and serve as a way of evaluating their actions and behaviour
- They have a role in the conduct of social life and help to build the norms to guide day-to-day behaviour
- give energy for doing something meaningful

2.1.2 How values are acquired

- From Parents, teachers, friends
- From groupings or settings such as the place of work, Religious grouping, learning institute
- From environmental interactions and influences
- From beliefs: values are derived and these can be correct or incorrect but still hold true for those who believe in them.

2.2 Ethics

Ethics are the set of rules or guidelines which govern behaviour and are usually established by a group or culture. They are derived from values. Ethics are implemented in the organisation to protect the interests of stakeholders (clients/customers, suppliers, employees, society and government. Thus they are needed to create conformity and order, and can enable members of a grouping (organisation, business) to interact harmoniously and are an enabler to achieving goals that would not be achieved individually.

Being ethical means supporting the realisation of positive values or the reduction of negative values [ISO/IEC/IEEE 42010:2011].

2.3 Why values and ethics matter

Values are the driving force in ethical decision making.

- They guide decision making. They represent viewpoints from which people make decisions
- They regulate our day-to-day behaviour and are a guide in setting objectives
- In the case of a company or organisation, ethics influence how the company or organisation works with and serves its stakeholders or stakeholder community. It is therefore expected that an organisation or company will have an ethical conduct which impacts on its processes, policies, procedures and supporting systems
- They help to set the boundaries between professional / business codes of ethics and personal codes of ethics.
- Organisations which practice ethics tend to be Agile for the following reasons:

- They will strive to provide excellence by working to deliver the greatest quality of service or products to the community / clients they serve. Such organisations will pursue creativity and innovation to deliver services and goods, and continue working to improve performance and stakeholder (client, employee etc) satisfaction and morale

- They will work to build and maintain positive reputation, keep stakeholders engaged (getting feedback on operational effectiveness)

- They take responsibility and accountability for the decisions they make and admit mistakes when these are made

- They are trustworthy, fulfilling commitments to stakeholders

- They show compassion and demonstrate kindness and care for others. This implies that decisions made seriously consider options put forward and how each option affects the other person or community with the aim of reducing the negative impacts or potential harms

- They manage risks(to achieve the above) to remain agile.

3. Incorporating values/ethics in IT systems and products

3.1 Why we should incorporate values/ethics into IT systems and products

Incorporating values / ethics into IT systems and products enables creators or developers to build products which better serve humans through functionality and behaviour which prioritise human values such as security, transparency, accountability, including the traditional organisational values such as efficiency, effectiveness and whatever values the organisation concerned subscribes to.

Although the significance of values and ethics is beginning to be realised more now and in the age of 4IR, the need to incorporate values / ethics in IT systems and products was realised some years ago. Notable early efforts include those made by Professor Mumford who developed an integrated methodology for systems implementation called Effective Technical and Human Implementation of Computer Systems (ETHICS) which incorporates job design as an important component of any systems planning and implementation effort [1]. Professor Mumford is known to have participated in the socio-technical movement which advocated for improvements of the quality of working life at the forefront of organisational design. ETHICS was viewed as a method of work design and change related to the introduction of computer systems. The methodology was developed to become a method of participatory systems implementation and is said to have evolved with each case study to suite an organisation's needs. It was popular because it promised management better success and gains, and workers better involvement and democracy in the workplace. Although there was no specific mention of terms such as values and ethics, the methodology's approach had focus on human values.

3.2 Some current methods of factoring or incorporating values into IT systems and products

In this chapter we will look at some of the existing and upcoming methods and frameworks which can be used to facilitate incorporation of ethics/values into IT systems and how these methods encourage innovation. A proposed framework on how the individual frameworks can be integrated to drive innovative development of IT systems, products and services will be also presented and discussed.

The importance of values in IT Systems, Products and Services started receiving attention in recent years with one of the key drivers being the Fourth Industrial Revolution (4IR) and AI. In the past, the focus was on requirements for economic gains and achieving operational efficiency with little or no attention to other values. The following are some of the methods:

3.2.1 Through corporate and IT governance frameworks

The IT Governance framework has matured over the years and is continuously being improved to meet the requirements of the modern organisation and the ever changing technology solutions delivery methods. The framework is modular and flexible and can be adapted to meet the requirements of organisations of different sizes. Typically, the governance framework delegates responsibility to the Executives to ensure that the enterprise's IT sustains and extends the organisation's strategies and objectives. ISO/IEC 38500 [2], the international standard for corporate governance of IT is more specific and identifies six principles for good IT governance which are as follows:

- Responsibility– the need to establish clearly understood responsibilities for IT
- Strategy – the need to plan IT to best support the organisation's business
- Acquisition – the need to acquire IT validly
- Performance – the need to ensure that IT performs well whenever required
- Conformance – the need to ensure that IT conforms with formal rules and regulations
- Human Behaviour – the need to ensure that IT respects human factors, through IT policies, practices and decisions

By implementing the above principles, the framework caters very well for factoring values into the organisation's structures and systems. The framework has been adapted and applied in public and private sectors and has proven to be effective in implementing the above principles to achieve the following key objectives regarding corporate governance of IT:

- Unlocking the value of IT. In addition to its traditional role of supporting business, IT is increasingly becoming an enabler and innovator of business
- Regulatory and Compliance issues which are changing rapidly. It is increasingly becoming important for organisations to comply, for example, to Protection of Personal Information
- Facilitating the acceptance by the Strategic Leadership of IT as an enabler of business
- Facilitating the provision of relevant resources, organisational structures, capacity and capability to enable IT delivery

- Implementing governance of IT based on known international standards and practices
- Appropriately empower and supporting IT leadership to ensure efficient delivery of the IT function
- Complying with Acts of Law.

Figure 1 is illustration of the adaptation of the framework which has been effective in both public and private entities.

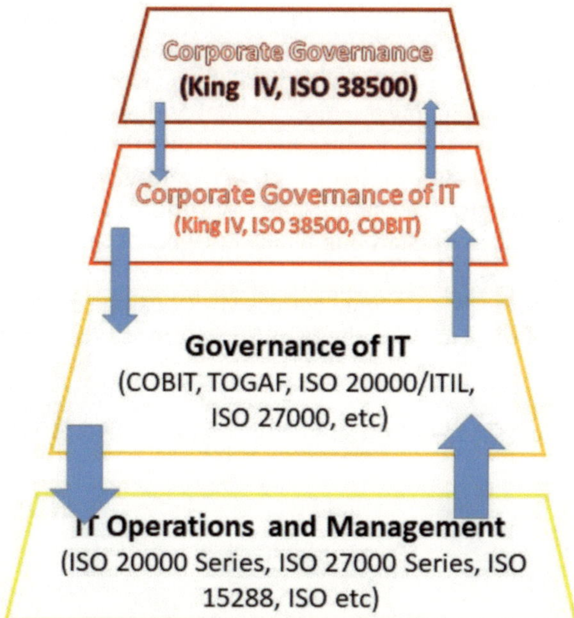

Figure 1.
Illustration of corporate IT governance framework implementation.

3.2.2 Ethically aligned design, development, management and implementation of IT systems, services and products

There are currently a number frameworks and methods some of which have already been released, and others in their advanced stages of development. Although the frameworks have been designed and developed with Autonomous Intelligent Systems (AI/S) in mind, they can all be applied to semi-autonomous and other types of systems, products and services. In a way, they all come in as a response to early calls by the likes of Mumford and others [1] for incorporation of ethics into the development of IT systems and products. The key drivers to these frameworks are an Ethical or Values based approach to the entire Systems Engineering and Systems Development Processes, and of Data Engineering and Management

- **Ethical Value Based Requirements Engineering(evb-RE) for systems/ products/services:** The Systems Engineering Body Of Knowledge (SEBoK) states that "System Requirements are all of the requirements at the system level that describe the functions which the system as a whole should fulfil to satisfy the stakeholder and requirements, and are expressed in an appropriate combination of textual statements, views and non-functional requirements, the latter expressing the levels of safety, security, reliability etc that will be necessary". The SEBoK further states that System requirements play major roles in systems engineering in that they form the basis of system architecture and design activities, form the basis of system integration and verification activities, act as reference for validation and stakeholder acceptance and provide a means of communication between the various technical staff that interact throughout the project. Incorporating Ethics/Values into Requirements elicitation adds another layer to traditional System Requirements [3] by incorporating ethics / values elicitation activities relevant to the domain culture and the surrounding environment where the system is to be deployed into a step which necessitates the introduction of additional activities or work items and skills requirements in the organisation. It further requires the involvement of a wider audience for quality purposes, dedication to human values. Falling short of using the term values, Bergman, et al., 2002, argues that effective requirements engineering brings together technical, social, economic and institutional factors, an approach which broadens the understanding which one gets of the domain of the System of Interest and helps to improve the effectiveness of the Requirements Engineering process [4]. The concept of evb-RE can be applied to Requirements Management in the TOGAF Enterprise Architecture Framework [5].

Figure 2 is an illustration of how evb-BRE can be applied to TOGAF.

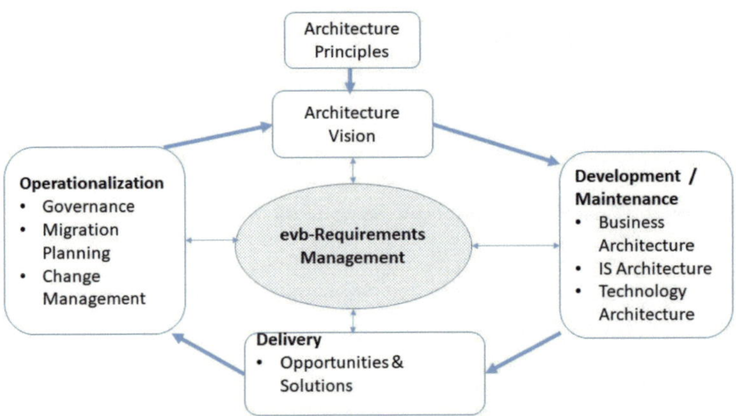

Figure 2.
Illustration of how ethical values based requirements engineering could be applied to TOGAF.

- **Ethical Risk Based Design**: The method works hand in hand with the Ethical Values Based Requirements Engineering [4]. It works by declaring identified values as values at risk (Ethical Value Risks) during the proposed system development effort. The identified values risks are treated through a repetitive and rigorous design process which aims to eliminate or minimise the risk to acceptable levels. Risk based design focuses on identifying stakeholders' attitudes and feelings about products and prototypes as an aid towards refining requirements with human centred values.

3.2.3 Using an appropriately designed well-being impact assessment framework to incorporate values into organisational processes, systems(including data), products and services requirements management systems

The proposed framework defines well-being as the continuous and sustainable physical, mental, and social flourishing of individuals, communities and populations where their economic needs are cared for within a thriving ecological environment. Through an iterative well-being impact assessment (WIA) process during conceptualization, analysis, design, development, and throughout the life span of a system, product or service the method establishes values of various stakeholders including creators/developers and uses the assessment findings to develop and improve a product, service or system of interest. It makes use of known knowledge bases or databases of values and the indicators of how those values can be impacted by the proposed system, service or product. New values and indicators can be identified during well-being impact assessment processes and these new findings are added to the database thereby continuously improving its content.

The Well-being Impact Assessment framework can be applied to any process and at any stage or phase in the life cycle of a system, product or service. However, in the case of systems, services and products it is strongly recommended that it be applied right from the beginning or start - at ideation, right through specification, development, testing, implementation, use, management and decommissioning. In this context, the use of the Well-being Impact Assessment Framework may be illustrated as follows:

- At the Governance and Executive Levels the WIA framework helps to initiate discussions by providing knowledge (to initiate, drive and guide discussions) from the information derived from reference databases (e.g. OECD)

- During ideation, the impact assessment process can initiate the requirements discovery and in this way makes contributions to comprehensive innovation management and ideation processes

- During system, product or service requirements elicitation and specification processes, it widens scope and coverage encouraging participation and contributions from all stakeholders

- During development, it opens up for a continuous engagement process between system/product/service creators or developers and all stakeholders. Values and related requirements are continuously reviewed and refined, and used to comprehensively define sprints in development

- During testing the framework is an enabler for collaborative and all-inclusive testing processes

- During live use and operation, the framework further opens up for collaborative and all inclusive system/product/service improvement processes. It strengthens and puts transparency into failure management processes

- At the decommissioning or retirement phase, the framework opens up for detailed engagements on user(all levels) experiences and performance levels achieved

- from all of the above, the values reference databases are continuously updated with information which is used to continuously drive innovative thinking in building and improving products, systems and services.

Whilst currently there is no data to support the results of this framework, the theories and features the framework promotes among which are cultures of continuous engagement and collaboration between stakeholders at all levels (legislators, policy makers, business executives, subject matter experts, creators of systems, products and services) point to increased coverage and levels of activity in the ethical values based (EVB) creation, use and management of IT systems, products and services. All these put a lot of pressure on requirements for innovative thinking. The traditional IT Governance framework has tended to evolve mainly from discoveries at the operational and lower levels, and never sounded loud enough in advocating for values. The framework outlined above gives the push at all and from all levels - again putting pressure on a requirement for more innovative thinking at all levels.

An example reference framework is IEEE 7010–2020 Recommended Practice for Assessing the Impact of Autonomous and Intelligent Systems on Human Well-being [6]. It is illustrated in **Figure 3**. **Figure 4** illustrates how it would interface with Corporate Governance and its supporting and implementation Frameworks.

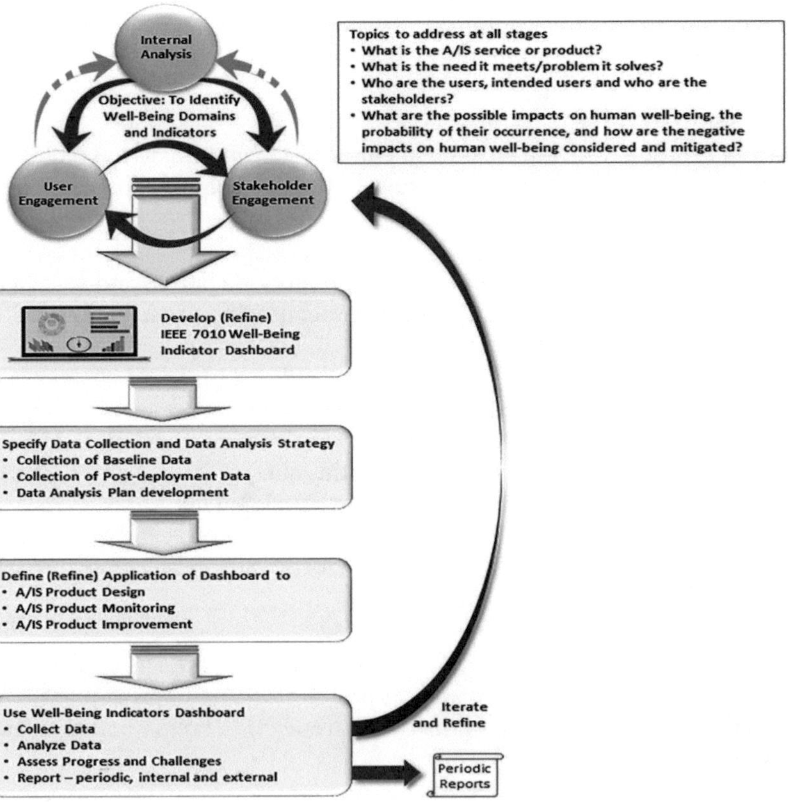

Figure 3.
The IEEE 7010-2020 well-being impact assessment framework (adapted and reprinted with permission from IEEE. Copyright IEEE 2020. All rights reserved).

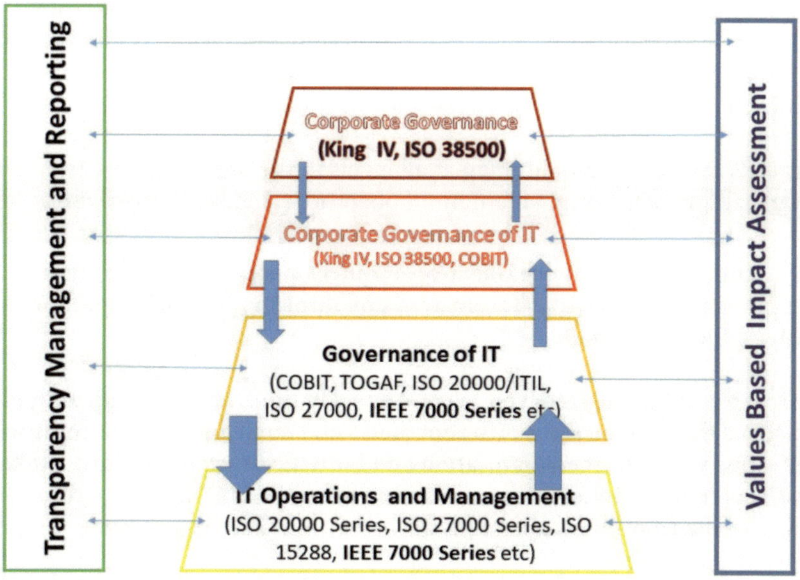

Figure 4.
Illustration of how impact assessment driven values based design can be implemented trough the IT corporate governance framework.

3.3 Potential challenges with incorporating values/ethics into IT systems, products and services

The following are some of the challenges that can be encountered when working with values

- People rarely express their values directly
- People rarely speak directly about their emotions. They illustrate their emotions through stories or speech patterns
- Values seem to be tacit knowledge – they are recognised when they are encountered. Therefore trying to articulate them before hand is difficult
- Values can be interpreted as a set of issues/value clusters e.g. consequences of automation, conflicts between stakeholders

Dealing with the above challenges requires a lot of effort coupled with creativity

4. Important observations

4.1 Values are a stimulus to thinking and therefore tend to motivate

Today and the future are about value creation. There is focus on what values a system or product brings into a setting or organisation.

4.2 Organisations which practice and value ethics continuously work to rediscover themselves so as to maintain their reputation as an organisation, and the reputation and quality of the organisation and services and products

This calls for innovative thinking at all levels in the organisation – from strategy to policy making, policy implementation, operations and service delivery.

4.3 Practicing ethics in an organisation requires people from different backgrounds and with different ways of thinking to work together to achieve the same goals

There is a need for all involved to continuously think and come up with ideas on how to strike the balance from different views and come up with concepts which bring positive values to the organisation and both its direct and indirect stakeholders. This requires innovative minds from all in the organisation – from the strategists to the frontline workers [4, 7].

4.4 An organisation known by its clients and stakeholders stands a very high chance of getting honest and valuable feedback on its service delivery levels from those clients and stakeholders

This opens up the organisation and its systems to continuous improvement and enhancements which require innovative minds.

4.5 Organisations design and implement processes and procedures to facilitate implementation of their mandate and delivery of their services

In fact an organisation's ideas and values are reflected in the way these processes and procedures are delivered. With innovative minds required to continuously rediscover the organisation through review of values in response to the environment, innovation is required to review and redesign the processes which support the ideas and values – striking the balance whenever this is required.

4.6 At the heart of process implementation, and services and products the processes produce are IT and related services and products

As an organisation continuously strives to rediscover itself to achieve excellence through implementation of its values, so should the organisation's support systems such as IT.

Incorporating values into systems / products/services can affect how teams for systems / product development are assembled and equipped, and how processes to deliver those systems/ products are structured. The following are example value concepts identified for a system/product and how incorporating them would affect the Requirements Engineering Process of producing the system/product

- Value Name/ID: Aesthetics e.g. Beauty, Presentation Potential source(s): Reaction to user interface of a product or system Implications on Requirements Engineering: Team member composition e.g. ensure you have UI designers.

- Value Name/ID: Security e.g. Safety, Privacy Potential source(s): Data Privacy laws, Data management policies Implications on Requirements Engineering

Processes: The process must include a threat analysis activity and must have cyber security experts as part of the project team.

4.7 The significance of value based systems engineering processes (described in paragraph 3)

The processes are designed to consider values, motivations and emotions. With emotions defined as responses to events, objects and artefacts, negative emotions (from stakeholders) can hinder system acceptance and use. Systems and software development have the ability to change working circumstances and in this way have an emotional effect on some stakeholders [4]. Therefore in order to succeed, systems, products and services are required to observe stakeholder values. For example if a system/product/service changes the power balance, it is important to:

- Consider political and social issues
- Consider stakeholders' potential emotional reactions to system change. These can be a major source of conflict hence system rejection and conflict e.g. stakeholder values of ownership and control
- Consider personal beliefs
- Consider conflicts between stakeholders' values and motivations and solutions proposed by requirements analysts
- Elicit stakeholders' attitudes to potential systems recognizing that:
    - Systems support decision making (Autonomous and Semi-Autonomous)
    - Systems and technology can change power balances in that they can take away or reassign responsibilities in a way which can strip or add powers to role players

All of the above will thrive in an environment where there is continuous engagement and innovative sharing and implementation of ideas from all stakeholders – including service/product consumers or customers.

4.8 Values considerations and Agile

It is believed that Agile was born out of a desire to create value. Where values come to play, there is agility and innovation. Responsive organisations are agile.

5. Recommendations

For a total sum, relevant organisational policy and policy implementation models and frameworks will need some form of integration or interface so that there can be an organisation-wide integrated response to the call to consider and incorporate values in the organisation's brand. Implementation of a suitable Corporate Governance Framework such as the one illustrated in **Figure 4** of which IT is a direct component is important and would be the key driver of this integrated response. This can drive and makes it possible for an organisation which adopts and implements such or similar framework to experience innovation resulting from the introduction or incorporation of values across its operational/business units and systems. Such a governance framework will guide implementation of operational frameworks such as Ethical Values Elicitation, Ethical Requirements Elicitation and Ethical Risk Based Design as discussed in paragraph 3, and a suitable Failure Safety Management Programme. In **Figure 5**, is an illustration of a proposed integrated framework showing the pillars for ethical values based IT systems, products and services and how all the frameworks discussed earlier work together to drive and support innovation to achieve the desired goals. It is designed to be adaptable and will target to consist of the following at a minimum:

- A values-based Corporate and IT Governance Framework
- A values-based Innovation Management Framework [8]
- A values-based Solutions Delivery Framework
- A values-based Well-being Impact Assessment or Monitoring Framework: This is an enabler of continuous review, learning and improvement of systems, products and services [6, 9]
- A values-based Operational Risk
- A values-based Failure Safety Management Framework.

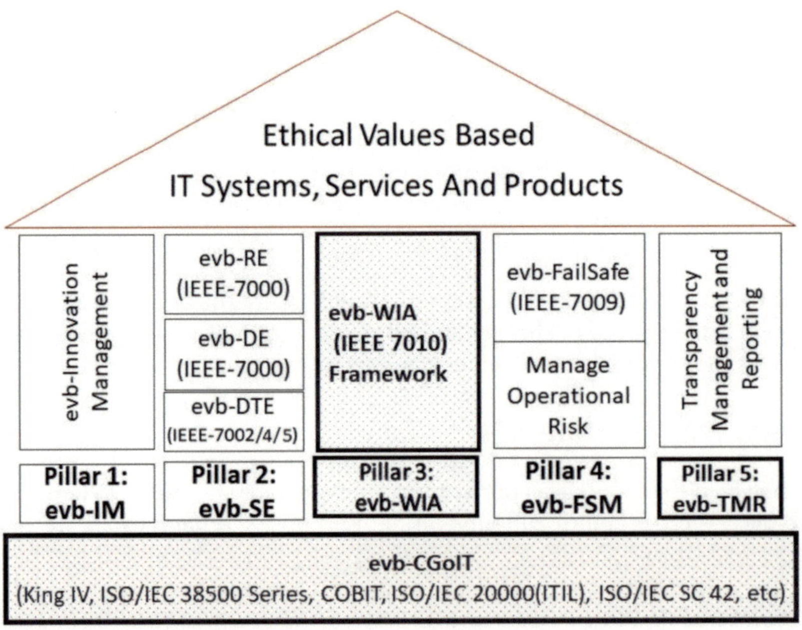

Figure 5.
The pillars of innovative ethical values based IT systems, products and services.

The position of the Impact Assessment Framework pillar is indicative of its central role as a monitor and assessor of the performance of the other frameworks and also as the key driver of continuous learning, innovation and improvement of systems, products and services.

Figure 6 is an illustration of how Impact assessment and values based design could be applied to TOGAF 9.

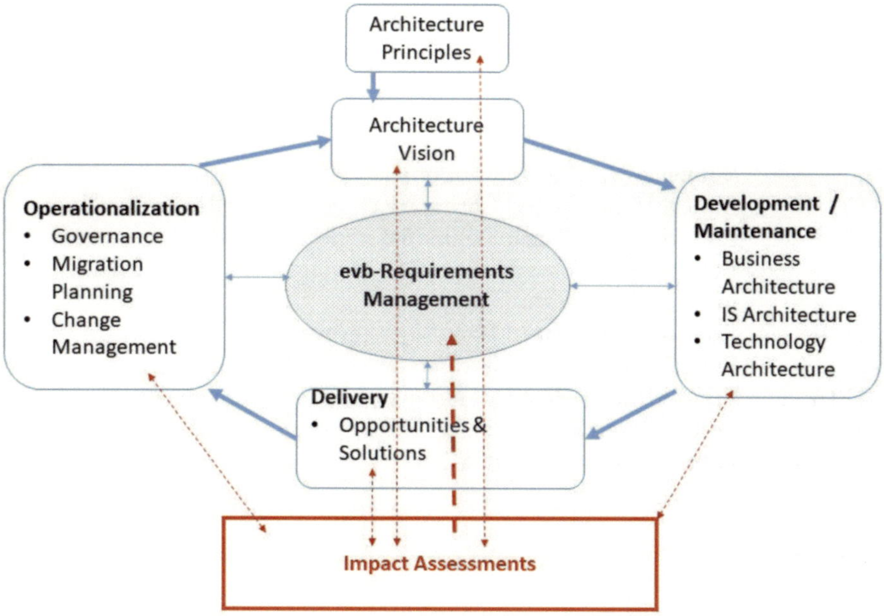

Figure 6.
Illustration of how impact assessment and values based design can be applied to TOGAF.

6. Conclusion

1. Ethics Considerations drive innovative thinking in building systems which are responsive to the needs of the user and promote adoption, and safe use of IT. It is when our most deeply held values such as safety, security, transparency, accountability, well-being are at stake that ingenious new solutions are needed most. It is therefore right to conclude that ethics are a driver of innovation and a source of competitive advantage.

2. Ethical considerations and shared moral values can be used to shape the world of tomorrow and should be construed as stimulus and opportunities for innovation, and not impediments and barriers.

3. Stakeholders' VME(values, motivations and emotions) can have significant impact on the outcome of a project. Innovative thinking is required to maximise positive impacts and minimise or even eliminate the negative(s)

4. It is useful to understand and have awareness of values and emotions as a guide and managing tool to the requirements analysis and elicitation processes when building systems and IT products.

5. Value Analysis may alert the analyst to potential stakeholder conflicts, when negotiations are needed to reach a common set of values. E.g. System configurations / customizations may need to be considered or when different levels of security controls mapped to stakeholders who regard security as very or not important. Balancing the act to make everyone happy requires innovative thinking.

Conflict of interest

The author declares no conflict of interest.

Nomenclature

CGoIT	Corporate Governance of IT
GoIT	Governance of IT
evb	Ethical Values Based
VME	Valiues Motivations Emotions
evb-CGoIT	Ethical Values Based Corporate Governance of IT
evb-FSM	Ethical Values Based Failure Safety Management
evb-IM	Ethical values Based Innovation Management
evb-DE	Ethical Values Based Design
evb-DTE	Ethical Values Based Data Engineering
evb-SE	Ethical Values Based Systems Engineering
evb-WIA	Ethical Values Based Well-being Impact Assessment
VBRE	Values Based Requirements Engineering
4IR	Fourth Industrial Revolution

Abbreviations

CGoIT	Corporate Governance of IT
GoIT	Governance of IT
evb	Ethical Values Based
VME	Valiues Motivations Emotions
evb-CGoIT	Ethical Values Based Corporate Governance of IT
evb-FSM	Ethical Values Based Failure Safety Management
evb-IM	Ethical values Based Innovation Management
evb-DE	Ethical Values Based Design
evb-DTE	Ethical Values Based Data Engineering
evb-SE	Ethical Values Based Systems Engineering
evb-WIA	Ethical Values Based Well-being Impact Assessment
VBRE	Values Based Requirements Engineering
4IR	Fourth Industrial Revolution

Author details

Zvikomborero Murahwi
School of Engineering and the Built Environment, University of Johannesburg, South Africa

*Address all correspondence to: zviko.murahwi@ictprojectsadvisory.com

References

[1] Elbana A, Newman M. The rise and decline of the ETHICS methodology of systems implementation: Lessons for IS Research. Journal of Information Technology. 2013: 124-136

[2] ISO / IEC 38500. Information Technology: Corporate Governance of Information Technology. ISO/ IEC. 2015.

[3] Hares AS. SSADM Version 4 - The Advanced Practitioner's Guide. Chichester, New York, Brisbane, Toronto, Singapore: John Wiley and Sons; 1994. ISBN: 0 471 93564 6

[4] Thew S, Sutcliffe A. Value-based Requirements Engineering: Method and Experience. Journal of Requirements Engineering. 2018

[5] The Open Group. TOGAF Version 9. Wilco, Amersfoort-NL: van Haren Publishing; 2009. ISBN: 978 90 8753 230 7

[6] IEEE 7010-2020. IEEE Recommended Practice for Assessing the Impact of Autonomous and Intelligent Systems on Human Well-Being. IEEE Standards Association. March 2020.

[7] Bednar K, Spiekermann. On the power of ethics: How value based thinking fosters creative and sustainable IT innovation. In: WU Working Paper Series. 2020. Available from: https: // epub.wu.ac.at/7841/

[8] Cohn S. A Firm-Level Innovation Management Framework and Assessment Tool for increasing competitiveness. Technology Innovation Management Review. October 2013

[9] A systematic methodology for privacy impact assessments: A design approach. European Journal Of Information Systems 23(2): 126-150.

Section 3

Ethical Education

Chapter 7

A Social Platform for Fostering Ethical Education through Role-Playing

Claudio Alvarez, Gustavo Zurita, Beatriz Hasbún, Sergio Peñafiel and Álvaro Pezoa

Abstract

Nowadays the complexity of knowledge, the specialization of labor and the pervasiveness of ICT in human activity, lead individuals to frequently make complex decisions with ethical implications. The educational system has a fundamental role in preparing specialized human capital in every discipline, however, it also faces the challenge of educating individuals with ethical discernment capabilities and behavior. In this book chapter, we describe the design, implementation and validation of EthicApp-RP, a social platform aimed at higher education settings, for fostering reflection and moral reasoning around ethical cases through a role-playing activity. We present an application of EthicApp-RP involving a cohort of undergraduate business students (N = 85), based on a case in which students play political and public leadership roles in the midst of the COVID-19 crisis. The results indicate that students and teachers acknowledge the learning environment's capacity to stimulate reflection and argumentation around ethical issues, while providing all students with equal opportunities for participation. In addition, the tool offers high technical and pedagogical usability, based on the Systems Usability Scale and the Pedagogically Meaningful Learning Questionnaire. EthicApp-RP can contribute to the improvement of ethics education, especially in scientific and technological disciplines, wherein students are quantitatively inclined by nature, in spite that ethics, a humanistic subject often foreign to them, must live at the core of their preparation.

Keywords: ethics education, higher education, human capital, social platform, educational technology

1. Introduction

Sometimes, people's behavior falls into unethical situations. Such behaviors are dependent on the context in which they occur, the points of view of those involved, the social norms in which people are framed, and what is considered morally correct [1]. In the world, several infamous cases of unethical conduct have come to light in academic [2, 3], governmental [4, 5], or corporate [6] contexts, among others. For example, in the Chilean national context, there are cases of collusion where various institutions have been involved, such as pharmaceutical corporations, radio stations, food companies, paper product companies, medical doctors, airlines, supermarkets, public transportation, etc. [7]. Internationally, one of the most notorious and recent cases of unethical professional conduct is that of the Cambridge Analytica scandal [8], which adds to dozens of other corruption cases that have occurred in different parts of the world [9]. Likewise, the impact on the environment, and technological advances in areas such as machine learning, cybersecurity and big-data, have generated new ethical dilemmas and situations in which professionals are expected to be able to deal with ethically [10].

One way to minimize breaches of ethical behavior involves incorporating ethics education into higher education, so that instructional activities and learning environments are provided, with the capacity to stimulate reflection, argumentation, ethical discernment and moral reasoning around ethical issues. In addition, it is of utmost importance that these opportunities equally reach all students in higher education, notwithstanding their gender, cultural background or whether their field of study is in the sciences or in the humanities. Higher education institutions have become aware of the urgency and relevance of these skills [11–15], considering them an essential and transversal component of academic curricula [16].

The literature identifies a growing need and relevance of ethics education in various learning domains and occupations, such as information systems [17, 18], auditing [19], marketing [20], taxes [21], among others. In the domains of computer science and software engineering, the software industry has been faced with an ethical crisis [22, 23], as users are increasingly aware about their personal data being utilized by platforms and services for various uses, including production of discriminatory profiles [24], and disinformation and fake news through massive manipulation of public speech, which has included electoral interference [25].

Professional and academic associations, as well as accreditation boards and agencies in a variety of fields, including engineering [26], computer science [27], business [28] and accounting [29] have taken notice of the importance of ethics in higher educational curricula and professional practice. In accreditation, there has been an increasing demand for the inclusion of courses in the areas of 'social, ethical and professional issues'. For instance, according to ABET [26], accredited computer science programs must cultivate "an understanding of professional, ethical, legal, security and social issues and responsibilities".

Although ethics is nowadays present in business and engineering school curricula, and it is part of the competencies in many of undergraduate and graduate profiles in universities around the world, teaching ethics in business [13, 30] or engineering [14, 15] is not a simple task, since there are epistemological, methodological and pedagogical differences in how teachers and students perceive ethics. Although there is a growing consensus that ethics teaching is important, there is little consensus on how to do it. Traditional forms of ethical training, including lecturing and case-based analysis dominate classrooms. These offer limited possibilities for students' active participation, which is highly desirable in their ethical training, [16, 31]. The activities in ethical training promoting the socialization of points of view, participation in discussions, reflection, and the development of ethical discernment are not those that predominate in traditional pedagogy. The ethical debates around ethical cases or dilemmas assume that students must not only demonstrate the ability to apply moral reasoning and develop ethical judgments, but also to communicate these processes competently and meaningfully, expecting to be heard, understood and respected by their classmates and the teacher [29].

2. Research context

With the intent of fostering the development of ethics skills in higher education, in the period 2018–2019 the present authors developed a collaborative web application called EthicApp, compatible with any current desktop computer or mobile device, including smartphones. EthicApp supports teachers in preparing and executing pedagogical tasks involving students' ethical discernment and reflection around ethical cases, in either face-to-face or online settings [32, 33]. Our early research with EthicApp focused on promoting higher level thinking processes, including reflection, argumentation, ethical discernment and moral reasoning. In addition, with EthicApp we strove to provide students equal opportunities for participation in ethics classes.

The first version of EthicApp consisted of a pedagogical flow comprising successive phases in which the students conduct ethical judgments individually, then in a small groups. The design sought that students express their judgments without inhibitions, so interactions among students were kept anonymous, even while working collaboratively. On the other hand, the teacher could monitor the activity and easily notice the groups of students presenting the greatest differences in the ethical evaluation of the case discussed. Lastly, the teacher could engage the entire class group in a discussion, for reflection on divergent ethical judgments found, and encourage students to further reason, argue and debate considering different points of view.

We conducted an initial pilot study of EthicApp, reported in [32], involving 35 Civil Engineering students from the Faculty of Engineering and Applied Sciences at Universidad de los Andes, Santiago, Chile. The analysis of students' behavior revealed that ethical judgments tend to be stable in the successive phases of the activity. However, it was observed that judgments tended to change more in groups where greater discussion occurred, and that the converse also happened. For this reason, we then considered that a desirable modification to the activity would consist in automating group composition, in such way that students with different views are brought together. Heterogeneous student grouping was thus hypothesized to increase students' interest in discussing the ethical case, and therefore, fostering a space where students have greater opportunity of modifying their ethical judgments as a result of argumentative and reflective processes in a social setting.

In [33], an experimental study was conducted with EthicApp in online mode, involving a cohort of 72 Civil Engineering students in the Professional Ethics Seminar course, in the same institution as in [32]. Greater chat interactions were observed among group peers in the heterogeneous grouping condition than in the random condition. In addition, it was identified, both in the heterogeneous and random grouping conditions, that the more chat messages were exchanged among the students, the more they produced argumentative discourse. Highly significant correlations were found among these variables. Lastly, it was found that male and female students benefit equally from the learning opportunities that are possible with the heterogeneous equally under the heterogeneous grouping condition, as no interaction effects were found among the quantity of chat message exchanges and gender.

In this chapter, we report on the development of EthicApp-RP, a social platform aimed at higher education settings, for fostering reflection and moral reasoning around ethical cases through a role-playing activity. In the following sections, we present the theoretical underpinnings of this research, the design principles of EthicApp-RP, the description of its instructional design, and a pilot study with business students to attest its technical and pedagogical usability, as well its effectiveness at fulfilling desirable qualities of role-playing activities in ethics education.

3. Theoretical and practical background

3.1 Ethics

According to [34, 35], ethics is a systematic approach to understand, analyze and distinguish issues of right and wrong, good and bad, admirable and deplorable in their relation to well-being and relationships between sentient beings. Ethics is an active process rather than a static one, which is why some ethicists use the expression 'doing ethics'. When people 'do ethics', they need to support their beliefs and claims with sound reasoning. In other words, even if people believe that ethics is totally subjective, they must be able to justify their positions before others through insights, reflections and arguments based on theory, context, rules, and rationality. In addition, feelings and emotions are a normal part of everyday life and can also play a legitimate role in ethics. However, people sometimes allow their emotions to outweigh good decisions related to ethics. Evaluations generated through the practice of ethics require a balance of emotion and reason. In contrast to ethics, morality is the set of beliefs, behaviors and specific ways of deriving from ethics. Morality can vary in a given population, depending on people's education, beliefs, social situation and culture. A person's morals are considered good or bad through systematic ethical discernment and reflection. The converse of morality is immorality, which means that a person's behavior is contrary to accepted social, religious, cultural or professional ethical standards and principles. Examples of immorality include dishonesty, fraud, murder, and acts of sexual abuse. Amoral is a term used to refer to actions that can normally be judged as moral or immoral, but which are performed with a lack of concern for good behavior. For example, murder is immoral, but if a person commits it without any feeling of remorse, or perhaps even a sense of pleasure, they act amorally.

According to [13, 35], ethics is a set of concerns, rules, principles, virtues, values and decision processes that allow people to live together and pursue their common and individual interests. As already made clear above, in the news everyday situations are seen that violate ethical principles in general, with a wide range of consequences for companies and citizens. Therefore, pedagogical artifacts and practices must be provided, which meet usability criteria [36], designed to help students from a wide variety of professionals, to be more ethical when making decisions in their future work fields. Ethical decision-making and moral reasoning are fundamental for future professional success and can be achieved by developing the skills of reflection, argumentation, discernments and moral reasoning, while students participate and communicate among them with equal opportunities [12, 37, 38].

3.2 Ethical discernment, reflection and argumentation

Ethical discernment is a characteristic of people that allows them to recognize the existence of an ethical dilemma, [20, 39]. The recognition of an ethical dilemma implies perceiving a problem or conflict in some situation or decision, whose dilemma becomes an ethical problem. It is considered that, if the ethical problem is not perceived, the process required to argue and reflect on ethical judgments will not happen [20, 39]. Therefore, it is necessary to distinguish both concepts: ethical dilemma and problem. An ethical dilemma exists when there is a situation where someone will consider one or more alternatives of action, including not acting at all, that are different consistent or inconsistent with some formal or informal rule, code or ethical norm [20]. An ethical problem does not exist until it is perceived as such, and then it happens that an attempt is made to resolve ethical dilemmas; that is, it occurs when a person perceives that their duties and responsibilities towards one group are inconsistent with their duties towards another group, including themselves. For the model developed by [20], only important ethical dilemmas will have an intense ethical conflict, provided that these are perceived as a type of problem. People often approach ethics with an initial expectation that there will be a correct answer to every question posed. It is important to help them accept the fact that there will not always be a correct answer, but one that requires personal judgment [40]. According to Kohlberg [29], ethical training should be encouraged, without limiting students to the role of mere spectators who only seek to apply the most appropriate ethical standard to each situation; but rather to carry out activities in which they participate, express, and make it easier for them to carry out their ethical evaluations, argued in an honest and authentic way, without feeling inhibited by their peers.

In [37, 40, 41], it is indicated that the key skills of ethical discernment are the following: a) **analytical skills** [41]: develop an aptitude for clear and logical thinking, where students learn to think reflectively, critically and solve complex problems supported by arguments that prove or refute the positions taken; b) **flexibility and independence of mind**: considering issues from multiple perspectives or points of view, encouraging a willingness to challenge orthodoxies, as well as the courage to set aside one's personal convictions to pursue a discussion wherever it leads; c) **making reasoned decisions** [41]: exercising coherent principles of thought and action, to learn to determine what types of evidence are needed to support their views and choices and that are justified by means of arguments that support the positions adopted; d) **communication skills**: learning to express points of view verbally and in writing, emphasizing group discussion and the articulation of arguments in direct response to verbal; and e) **group and collaborative work skills**, [37, 41]: create a supportive environment for the development of ethical discernment that is group and collaborative, where students feel safe, there is a climate of mutual respect and confidentiality is ensured. According to [40], the group and collaborative work skills required by a person who is dealing with an ethical dilemma are: a) share their ideas, either verbally or in writing; b) express their opinions without interruption; c) express their criticism, directed at arguments and not at individuals; d) be able to handle conflictive situations; e) encourage others to generate constructive criticism of their beliefs; f) encourage the search for commonalities between opposing points of view; and g) be open to considering different points of view.

3.3 Ethics in higher education

According to [35], there is a growing need for well-established ethical frameworks and practices in ethical training in business schools [13, 28, 42–44], and in engineering education [14, 15, 45, 46]; who have the responsibility of providing their students with training in their ethical discernment, argumentation and reflections [30]. According to [40], if a business or engineering schools provides what we call 'reactive' ethics education, which only serves to inform the practice of statutory and regulatory requirements and responsibility to shareholders, it is most likely that the organizations introduce procedures that merely comply with legal ethical business practice. However, for organizations to adopt an ethical stance and socially responsible thinking, they also need to be 'proactive', with fundamental ethics programs taught by business and engineering schools. A 'proactive' ethics education implies the development of flexible but ethical managerial thinking and practice, that can be applicable to different contexts. For this, it is necessary that business and engineering schools establish the importance of contributions that increase moral reasoning, the improvement of ethical training and the development of decision-making skills with an ethical approach and leadership. Boo and Koh's research [47] identifies that top management support, with links between ethical behavior, professional success, and ethical organizational climate are all that is necessary for effective ethical codes. It could be argued that corporate malfeasance, as in Enron and Parmalat [48], would have been prevented by properly employing and monitoring 'reactive' and 'proactive' business ethics practices and procedures. Clearly more profound changes in culture are necessary, and values are needed in organizations to deal with these types of problems.

According to Holsapple et al. [15], teachers in engineering schools often describe ethics education as a balance between knowledge of ethical codes of conduct and understanding of ethical rights and errors. However, graduates often report that their ethical training relied almost entirely on the application of codes, implying less depth and complexity in the analysis of ethical dilemmas. While ethics is intended to be a central component of today's engineering curriculum, it is often perceived as a marginal requirement that must be met [14]. According to [45], the pedagogy of ethics for engineers must consider the characteristics of thought inherent in the scientific training of students and their future professional approach. The authors characterize the mentality of engineers with the following description: the real world is what can be touched and measured, the prototype of rational thinking is mathematical-deductive reasoning, and the best results are obtained by following standard procedures. Therefore, it is a priority to recognize the difficulties of engineering students to recognize the value of ethics, along with moral discernment and reflection. In a systematic review of the literature on interventions for teaching engineering ethics in the USA, Hess and Fore [49] report that the most common methods involved exposing students to codes or standards, using case studies (cased-based learning) and discussions. They emphasize the need to develop learning experiences where students reflect on their own emotions and those of others, with greater empathy with the actors involved and the situations.

3.4 Instructional approaches in higher ethics education

In [37], a quantitative grouping procedure was carried out to derive a typology of instruction in ethics education with respect to four categories of instruction. These include content, processes, methods of delivery, and instructional activities. Eight instructional approaches were identified through this grouping procedure, each with different levels of effectiveness based on one of nine commonly used ethical criteria. Viable approaches to ethics training, of which effect size estimates (i.e., Cohen's d) are known, include 'professional decision processes training' ($d = 0.50$) and 'field-specific compliance training' ($d = 0.46$). Professional decision processes training uses a variety of techniques, including case-based learning, role-playing learning, problem-based learning, team-based learning and discussion. Next, articles that report on methods for developing ethical discernment and reflection in higher education contexts are described, comprising case-based learning, and role-playing activities.

Case-based learning (CBL) consists of the use of fictitious or real cases associated with specific curricular disciplines, in which ethical dilemmas are presented, and pedagogical activities of ethical discernment are instantiated. Students read and analyze a 'case' described in detail, usually adopting the role of decision makers [50]. Some examples of the use of this methodology are described below.

In the Faculty of Economics and Business at University of Chile, based on the contents of the cases described in [51], a methodology is applied based on a) case reading, b) identification of relevant actors, c) identification of premises in conflict, d) evaluation of alternatives and decision-making, and e) plenary discussion and conclusion in teams of 5 to 7 members, with the support of the socrative.com application to collect opinions. This methodology is applied in various courses requiring ethical education, such as Management and Business, Costs and Budgets, Business Income, Tax Economics, and Introduction to Economics, among others. Several advantages have been observed, including that the group discussion permits listening and analyzing diverse perspectives, improving the depth of analysis and discussion of the case and the ethical dilemmas identified. The moderator facilitates aspects to be debated and opens instances to spur students' critical reasoning. The use of socrative.com allows to have a record of the conclusions of the groups, thus facilitating the teacher's review after the session is finished. Among the cons, not all students' opinions can be effectively captured, because some students are apprehensive about openly exposing their comments. Moreover, limitations on the quantity and quality of the interactions arise due to time restrictions, and that the activity is done in a single class session.

In [52], to establish the case, face-to-face interviews with people directly or indirectly involved with business ethical dilemmas in real life are organized in class, so the experience of the actors involved is counted on. Then a discussion is held among the participants based on a specific ethical dilemma. The advantage of this variant is that, by being in contact with the person interviewed, it is possible to have a more direct contact with the various ethical dilemmas that are experienced in the professional field. It is expected that this level of proximity to the problem will allow the generation of greater affective empathy in the students and thereby improve their decision-making in real situations, taking advantage of the 'sensitization' of the students as a benefit of the process. In addition, while discussing during class, students learn from each other by presenting their own arguments that support the decision made. As for the disadvantages, it requires great preparation to be carried out, since it implies counting on a person involved in a real case. A record of what was discussed with the interviewee is not generated, but only what was noted by the interviewers, so the teacher does not know the points discussed by the group at the time of generating the discussions. This methodology was applied to ethics courses at Kenan-Flagler Business School of the University of North Carolina at Chapel Hill.

The CBL is a useful method to bring students closer to real ethical and professional decisions, without the consequences that decision making entails for case roles and stakeholders in real situations. The method manages to generate both the capacity for critical analysis and cognitive empathy. Despite being a method with many advantages, its main disadvantage is its structuredness. Preparation of structured cases is required, with sufficient contextualization and depth to understand the problem and achieve a connection between the students and the roles. In contrast, real life scenarios are often ill-structured and decision making relies on limited information.

Role-playing Learning (RPL) Role-playing is the exercise of changing one's behavior to take on a particular role. For this purpose, it is a conscious change to represent an adopted role, extracted from a context or problematic in analysis. It is a method that is regularly combined with the roles of people who are described in a case. Some examples of the use of this methodology are described below.

The method used in the York University School of Engineering [46] applies role-playing with theatrical elements to teach decision making on controversial ethical issues. The activity encompasses the following phases: a) role assignment: each student receives information on their role based on a script prepared by the teacher; b) discussion: the teacher presents questions about the case, and each student exposes and discusses their points of view, based on their role, in relation to these questions with their classmates; c) deepening: at some point in the discussion, the students can elaborate more detail about their positions, and expect their classmates to do so as well. With additional details provided by each role, the students can complete their analysis of the situation; d) plenary: after the discussions, the teacher begins a closing phase, in which the analyses achieved previously for each question are synthesized. Theatrical elements are used through the role-playing process, including costumes, music and other recorded media, which allow to further increase the credibility of the recreation. The authors who propose this activity [46] indicate advantages compared to other traditional methods, similar to those found other RPL designs, such as greater student involvement, engagement and dynamism in interactions. Among the disadvantages, it is indicated that a high degree of preparation is required, including the activity script, the description of the roles involved, as well as the theatrical resources that complement the exercise.

According to [43], who proposed an RPL activity that was incorporated into a financial management course for undergraduate and graduate students, RPL has the advantage of creating low-risk conditions so that students can express their opinions and perceptions with minimal teacher intervention. For RPL to be successful, the activity needs to be potentially conflictive, and ideally allow the majority of students to identify with some role, in order to encourage participation. The roles should result in personifications by which students can feel comfortable and immersed. Otherwise, the students will unlikely be able to imagine the actions the role would likely perform, nor relate their own experiences emphatically with the situation as experienced by the role.

The RPL is a dynamic and simple method to understand, and it allows to keep students more involved in the case or problem, since they internalize themselves from their role to defend their positions, and from where it is attractive to keep participating. As a general disadvantage, it is observed that identifying with a unique role in the game and defending their position from the perspective of that role, can cause students to then focus the solutions on the character they had to interpret, closing the possibility to the other characters or, sometimes, reducing the role of the decision-maker in the case.

4. Design of EthicApp-RP

4.1 Design principles

Based on the analysis of literature in the field of ethics education already exposed in [32, 33], this section present the design principles for EthicApp-RP, comprising relevant functions for supporting case-based learning in ethics [31, 53, 54] and role-playing [43, 44, 46]. Its design principles are as: 1) embeddable in traditional courses, 2) easy to use, 3) implicit interactions to support student and teacher roles, 4) multidimensional judgements, 5) anonymity, 6) support for reflection, discussion and argumentation, 7) domain independency, 8) efficient information management, 9) combine individual work and group work, 10) Flexibility, and 11) device independence.

All these principles are explained in detail in section 5.1 of a previous research which instructional design were based entirely on a cased-based learning methodology, and applying differentials to the selection of statements [32]. Regarding to requirement 2) a desirable level of technical usability is given by mean score equal or above 75 in the System Usability Scale (SUS) [36]. For requirement 4) applied to EthicApp-RP, the students must express their ethical judgement on the given case by ordering (i.e., prioritizing) a set of actions, according to a prescribed criterion, and by providing justification on the ordering of one or more of the actions. The criterion prompts the student to reason according to their assigned role, and based on that specific perspective, prioritize decisions considering their effect on different stakeholders, with the intent to reach the most beneficial (or least detrimental) solution pathway to the ethical problem. In order to attest the qualities of requirement 6), mean scores in the range of 4 to 5 points in the constructs of the Pedagogically Meaningful Learning Questionnaire (PMLQ) [55], are considered a desirable objective. Regarding 19), EthicApp-RP supports flexibility in its pedagogical flow; that is, while the activity must always begin with a mandatory individual phase to collect students' initial appraisal of the case, the successive phases, i.e., individual or collaborative, shall be optional and configured on-the-fly, thus allowing different phase configurations depending on timing constraints and pedagogical goals in which the activity is enacted, [37, 41].

4.2 Instructional Design of EthicApp-RP

The design of EthicApp-RP permits the teacher conducting role-playing activities comprising an arbitrary number of phases, including both individual and collaborative work. In spite of this flexibility, activities based on EthicApp-RP will commonly follow the jigsaw Collaborative Learning Flow Pattern [56]. Under this pattern, the activity is structured based on the following successive phases:

Prerequisites and Setup: To create and configure an activity, the teacher must set up its configuration, see **Figure 1(a)**. For this, they indicate its title, a brief description, and provide a PDF file containing the description of the case involved. In addition, the teacher defines a set of roles involved in the case, inputs a list of actions that the different roles must hierarchically order, enters the criteria by which the actions must be ordered by the students, each of them assuming a specific role. In addition, several other parameters can be configured by the teacher, including which actions require the students' written justification, whether the next phase of the activity is individual or collaborative, the type of groups that shall be formed (i.e., 'expert groups' or 'mixed groups'), and whether students' anonymity is required.

Individual Work: Each student reads the case presented and issues their first ethical assessment individually, according to their role. To carry out the ethical judgment, the student has to order the presented case actions according to the

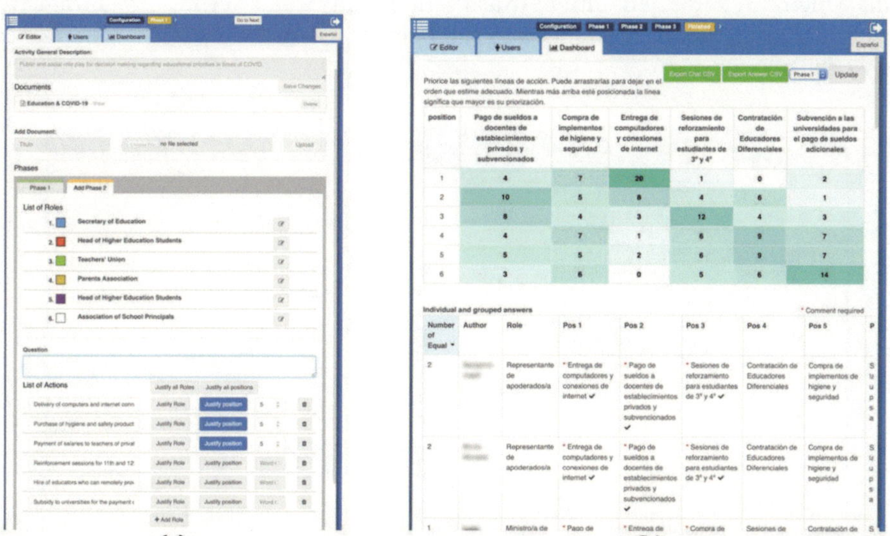

(a) (b)

Figure 1.
(a) Activity configuration panel, which allows the teacher to configure and start activity phases with different configuration parameters on the fly, (b) Teacher's progress dashboard, where students' and groups' progress can be seen.

required criteria, see **Figure 2(a)**. In addition, the student may need to provide written justification for the specific ordering of one or more actions, according to how the activity had been previously configured by the teacher. While the students work on this phase, the teacher can monitor their progress through a dashboard, see **Figure 1(b)**. The dashboard displays a matrix showing the frequency with which students place actions in the different orders that are possible. In addition, the teacher may see details of the response of any individual student.

Expert Groups: When the teacher transitions to this phase, EthicApp-RP implicitly groups students homogeneously, i.e., forms groups comprising students with the same role. The students discuss their prior individual responses, by means of anonymous text-based chat, see **Figure 2(b)**. They may re-elaborate their responses if they choose to do so or may maintain their response unchanged as in the previous phase. As in the previous phase, the teacher is presented with a dashboard through which they can monitor student' activity. The dashboard continues to present the matrix previously described, along with students' responses, and the possibility to see the groups' conversation through chat messages.

Mixed Groups: After the 'Expert Groups' phase finishes, students keep their role and EthicApp-RP forms groups composed of mixed roles. The number of students per group relates to the number of roles in the activity. EthicApp-RP's grouping algorithm attempts to form groups in such way that a single representative of each role is present in each group. Students in mixed groups must defend the interest of their assigned role, while at the same time pay respect to and consider their peers' different points of view. Like in the previous phase, students can modify their response after considering their peers' points of view and arguments.

Plenary Discussion: After the 'Mixed Groups' phase is over, the teacher can advance to a Whole Class Discussion phase, where they can present conflicting ethical judgments from different groups to the class and ask students to express their points of view and private assessments on the case. The teacher should be careful to select contradictory or divergent judgments judiciously to stimulate a discussion that will lead to an ethical based case resolution. The objective is for students to recognize the virtues of the resolution reached in this final discussion, which can help them build ethical schemes, as well as ethical meanings that they can transfer to different cases in their future as students or professionals in the workplace.

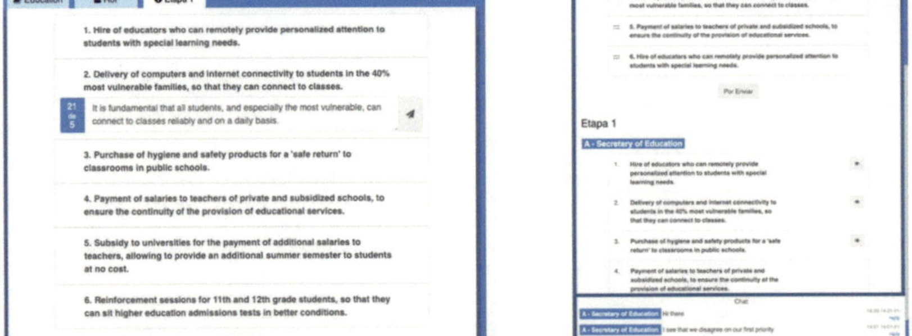

(a) (b)

Figure 2.
Students' user interface, showing (a) the Individual Response phase, in which the student ranks actions and provides justification for it, (b) the 'Expert Groups' phase, in which students with the same role discuss their responses anonymously.

5. Pilot study

'Social and Professional Environment' is a compulsory, first-year course, in the curricula of 'Information Engineering and Management Control', and 'Accounting and Auditing' degrees at the Faculty of Economics and Business ('Facultad de Economía y Negocios', FEN) at Universidad de Chile, Santiago, Chile. This course aims that students reflect on current socio-environmental challenges in relation to their future professional occupations. Since 2013, Ethical Discernment (ED) was integrated as a transversal skill at FEN, thus it ceased being taught as a dedicated course. The ED competence is defined at FEN as *"the use of a set of criteria that guide the projection of effects and consequences in decision-making in the field academic, professional and/or labor, considering norms, values and good practices"*.

The course is focused on five main themes: 1) Sustainable Human Development, 2) Poverty and Inequality, 3) Education, 4) Citizen Participation, and 5) Multiculturalism and Gender. Critical discussion is fostered based on these themes, for which students are provided the pertaining literature. In each course topic, special attention is paid to students' ability to analyze social problems and ethical issues raised, as well as establishing links with professional practice, and proposing possible solutions to the problems. Consistently with this rationale, the analysis of ethical dilemmas is part of the course methodology.

	T	Description (T: Time in Minutes, A: Asynchronous)
Before class	A	Students were announced that an ethical discernment activity would take place next class. They were asked to create their account at EthicApp-RP, and read the case text, available at the course website.
Briefing	5	The students were welcomed to class, and the objective of the activity was presented. A general summary of the case was then displayed.
Ethical Case Reading	5	The students entered EthicApp-RP with their credentials, and found the text of the case, so that they had it available during the activity.
Individual Work Phase	5	Each student was assigned one of the following roles: Secretary of Education, Secretary of Finance, Head of Higher Education Students, Parents Association, and Association of School Principals, Teachers' Union. Details about the assigned role and their participation in the case were provided as well. Each student had to individually adopt their role and prioritize the actions.
Expert Groups Phase	15	Groups of 4–5 students were formed, with all students having the same role. Each group was asked to first reflect on the case and the actions proposed. Then, students in their assigned roles were asked to take a position in the case, by prioritizing the actions, and providing justification for the chosen prioritization.
Mixed Groups Phase	15	New groups were formed, this time composed of students with different roles. Each student had to defend the interests associated with their role. Then each student was again asked to prioritize the lines of action, and to provide justification for the prioritization, considering the discussion that just took place in the mixed group.
Plenary Discussion	15	A final plenary session was held, where each heterogeneous group (i.e., from the previous phase) presented their prioritization to the rest of the class. The discussion emphasized the importance of considering the stakeholders of interest in the decision-making process, as well as the changes in prioritizations found through the three previous activity phases. Lastly, the students were asked to share their impressions and feedback on the activity through a survey at menti.com.

Table 1.
Description of the role-playing activity based on EthicApp-RP.

ID	Action	Description
A1	Hygiene & Security	Purchase of hygiene and safety products for a 'safe return' to classrooms in public schools.
A2	Special Educators	Hire of educators who can remotely provide personalized attention to students with special learning needs.
A3	Devices and Connectivity	Delivery of computers and internet connectivity to students in the 40% most vulnerable families, so that they can connect to classes.
A4	Teacher Salaries	Payment of salaries to teachers of private and subsidized schools, to ensure the continuity of the provision of educational services.
A5	11-12th Grade Tutors	Reinforcement sessions for 11th and 12th grade students, so that they can sit higher education admissions tests in better conditions.
A6	University Funding	Subsidy to universities for the payment of additional salaries to teachers, allowing to provide an additional summer semester to students at no cost.

Table 2.
List of actions prioritized by the different roles in the activity.

Due to the COVID19 pandemic in 2020, the development of the course was faced with the challenge of maintaining the active learning methods in an online format, as these were customary in face-to-face classes. For this reason, it was decided to pilot EthicApp-RP in the course, in order to facilitate conducting role-playing activities in the third course unit. An ad hoc ethical dilemma was written, based on the challenges that the country was experiencing due to the pandemic. The dilemma closely resembled the national reality at the time of the activity.

In total, 85 students participated in the trial activities, divided into two sections of 49 and 36 students, respectively. In both sections the activity lasted one hour, however, it was conducted at different times and guided by different teachers. Participation was entirely online, with use of Cisco Webex for synchronous communication.

5.1 Role-playing learning activity

Table 1 summarizes the steps followed in the pilot activity. Before class, the students had to study the case, which basically described the state of events in the Chilean education system in the midst of the COVID-19 crisis. In synthesis: The education system had been challenged with the need to migrate all levels of education to online formats. Adoption of online education meant that all educational levels had to sacrifice learning outcomes and contents, due to reduction of effective class time.

In the individual work phase, each student was assigned the role of a decision maker, automatically, by EthicApp-RP (see **Table 1**). According to the assigned role, each student had to prioritize a set of actions to cope with the crisis (see **Table 2**). The intent was that each student prioritized the actions considering resource limitations, and the interests of the stakeholders they represented and society as a whole. Next, the 'Expert Groups', 'Mixed Groups' and 'Plenary Discussion' phases ensued.

6. Quantitative results

The entirety of the cohort, i.e., 85 students, connected to EthicApp-RP at the beginning of the activity. However, two students entered late and were not assigned to a group, thus only 83 participated in the first phase, and 81 thereby submitted their response. In the role assignment performed by EthicApp-RP in phase 1 (i.e., 'Individual response'), there were between 13 and 15 students assigned to each role. The roles were assigned to the students in the following quantities: Secretary of Education to 15 students, Secretary of Finance to 14, Head of Higher Education Students to 14, Parents Association to 13, Association of School Principals to 13, and Teachers' Union to 14 students.

In phase 2 (i.e., 'Expert Groups'), 81 students participated. In the first section, the groups were more numerous, composed of 7 to 9 students, while in the second section, the groups involved from 5 to 7 students. Finally, in the third phase (i.e., 'Mixed Groups'), 80 students submitted responses.

Regarding chat messages, a significant increase was observed between phases 2 and 3 (see **Figure 3**-left), especially in the roles of Secretary of Education, Secretary of Finance, and Head of Higher Education Students. This is an expected behavior in mixed groups, since in previous studies it has been determined that in groups where there are different points of view, the discussion is greater than in groups with more homogeneous views [32].

The distributions of chat messages per student considering the different roles follow a similar trend to that observed with respect to the totality of messages by role (see **Figure 3**-right). In particular, in phase 3, certain outliers are observed for the roles of Secretary of Education and Secretary of Finance.

Through the successive phases of the activity, all roles placed action A3 – 'Devices and Connectivity' as the first priority, and the last priority was that of A6 – 'University Funding' (see **Figure 4**). It can be seen that the priorities evolved throughout the three phases; however, the first three priorities remained relatively stable. Apart from the first priority already mentioned, in second place of priorities, the action A4 – 'Teacher Salaries' dominated in the three phases, and in the third place, there was a similar number of preferences for A5 – '11-12th Grade Tutors', A4 – 'Teacher Salaries', and A2 – 'Special Educators'.

Figure 5 shows Sankey networks depicting how first priority preferences evolved in each of the roles through the three phases. It can be seen that the first priority varies according to each role. Notably, in the first phase, the first priority

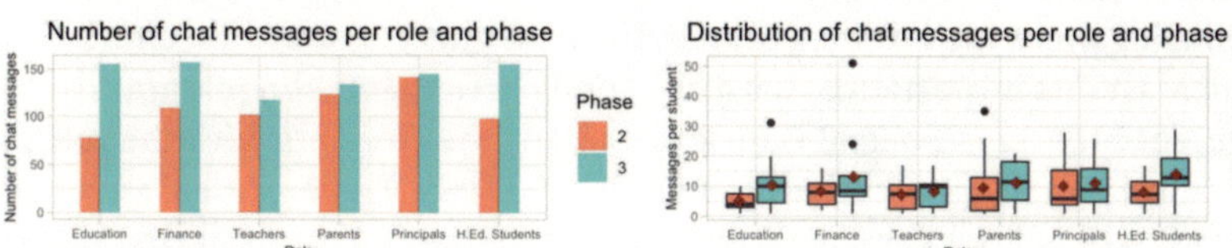

Figure 3.
At the left, number of chat messages per role and phase. At the right, distribution of chat messages per role and phase.

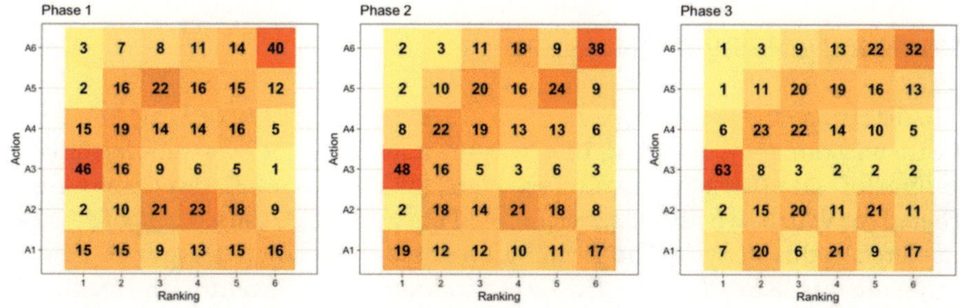

Figure 4.
Frequency of action rankings per phase.

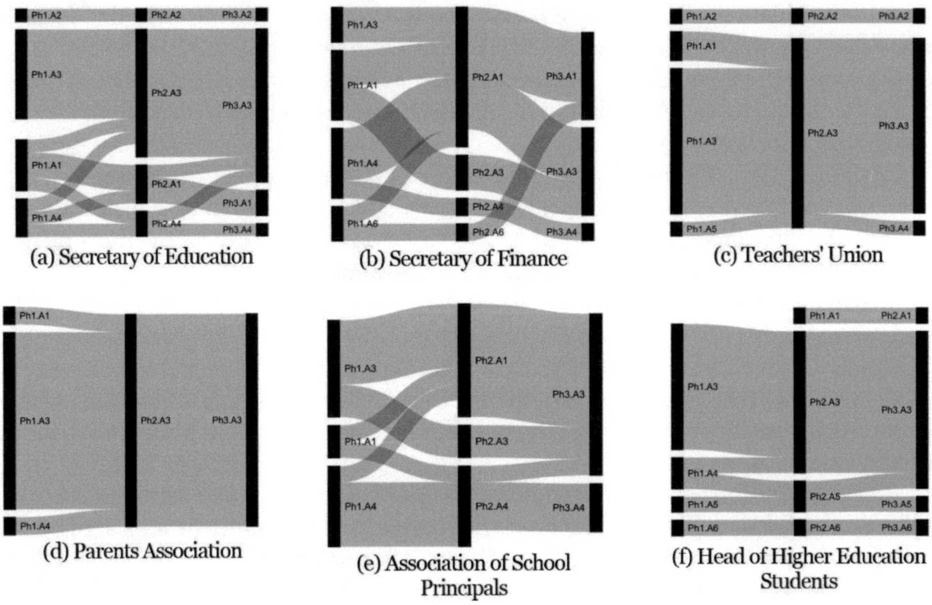
(a) Secretary of Education (b) Secretary of Finance (c) Teachers' Union (d) Parents Association (e) Association of School Principals (f) Head of Higher Education Students

Figure 5.
Sankey charts depicting the evolution of the first priority chosen by students with different roles throughout the activity.

for the secretaries of Education (a) and Finance (b) is distributed in four actions. In contrast, for Teachers' Union and Parents Association, the action A3 – 'Devices and Connectivity' clearly dominates. In the case of Principals, there are three priority actions, and in the case of the Head of Higher Education Students, action A3 is dominant, and three other actions have less weight.

In the second phase, of expert groups, the first priority was changed in each of the roles. In the case of the Secretary of Education, A3 starts to acquire major importance. In the case of the Secretary of Finance, A1 – 'Hygiene and Security' acquires much greater importance than in the first phase. For the Teacher's Union role, A3 increases its importance, with only one participant who maintained their preference for A2 – 'Special Educators'. All of the students with the role of Parents Association representatives prioritized A3 first. The representatives of the Association of School Principals maintained the same actions in first priority, increasing in importance A1. Finally, for Head of Higher Education Students, A3 increases its importance and a student appears prioritizing A1 first.

In the third phase, only the role of Secretary of Finance maintains a prioritization where A1 and A3 are equally divided in first place. For all other roles, action A3 takes on the highest importance.

6.1 Technical usability of EthicApp-RP

To determine the students' perception of technical usability of EthicApp-RP, the SUS questionnaire, based on 10 Likert 1–5 scale items [36] was administered in online format to the participating students, [57]. A total of 39 responses were gathered, of which two responses were ruled out as invalid, thus 37 responses are considered in this analysis. The distribution of scores is shown in **Figure 6**. The mean score was 78.6/100 (SD = 13.8), the minimum was 47.5 points, the median 75, and the maximum 100. Only four students (11% of responses) gave a score lower than 68, which is considered average usability according to [36].

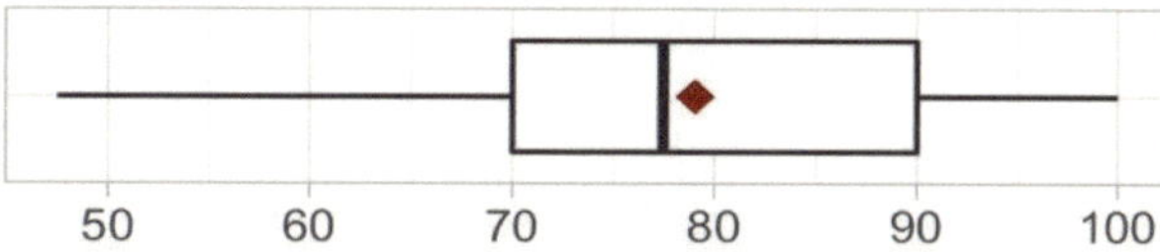

Figure 6.
Distribution of EthicApp-RP SUS scores.

N	Item	*M*	*SD*
1	I think that I would like to use EthicApp-RP frequently in university courses.	4.38	0.70
2	I found EthicApp-RP unnecessarily complex.	1.80	0.85
3	I thought EthicApp-RP was easy to use.	4.25	0.63
4	I think that I would need the support of a technical person to be able to use EthicApp-RP.	2.20	1.04
5	I found the various functions in EthicApp-RP were well integrated.	4.25	0.71
6	I thought there was too much inconsistency in EthicApp-RP.	1.88	0.79
7	I would imagine that most people would learn to use EthicApp-RP very quickly.	4.30	0.52
8	I found EthicApp-RP very cumbersome to use.	1.52	0.72
9	I felt very confident using EthicApp-RP.	4.03	0.80
10	I needed to learn a lot of things before I could get going with EthicApp-RP.	2.17	1.01

Table 3.
EthicApp-RP Usability Scale results.

Construct	Abbrev.	Items	*M*	*SD*
Applicability	AP	5	4.48	0.45
Added Value	AV	3	4.42	0.55
Cooperative/Collaborative Learning	CL	3	4.57	0.59
Feedback	FE	3	4.41	0.54
Flexibility	FL	4	4.44	0.50
Goal Orientation	GO	3	4.47	0.55
Learner Activity	LA	3	4.13	0.61
Learner Control	LC	4	4.35	0.60
Motivation	MO	3	4.53	0.45
Valuation of previous knowledge	VP	3	4.49	0.52

Table 4.
Results of the PMLQ instrument, by construct.

Table 3 shows the items from the SUS questionnaire, with their respective descriptive statistics. Students consider the use of the relevant tool in university courses (question 1). On the other hand, it is observed that the application was, on average, easy to learn to use (question 3), without the students having received training prior to the activity carried out. Also, the functions are easy to remember (question 4) and understand (question 5) for most students.

6.2 Pedagogical usability of EthicApp-RP

The evaluation of Pedagogical Usability, that is, the appraisal of the pedagogical qualities of the design and the use of EthicApp-RP, was carried out through an adaptation of the Pedagogically Meaningful Learning Questionnaire (PMLQ) [55], with a total of 34 Likert items in a 1–5 scale, [58]. This instrument allows evaluating pedagogical usability considering a series of criteria, as shown in **Table 4**. PMLQ was applied in conjunction with SUS, hence the same number of valid responses was obtained, i.e., 37 out of a total of 39.

It is observed that in every construct the average score obtained is in within the range of 4–5, which meets the pedagogical usability goals established at the outset of EthicApp-RP's development process.

7. Qualitative results

Students' written justifications in the EthicApp-RP activity were analyzed for complexity of ethical reasoning. For this, the rubric of the Association of American Colleges and Universities (AAC&U) on Ethical Reasoning was used [59]. This rubric was preferred, since it was developed by teams of faculty experts representing colleges and universities across the United States, "*through a process that examined many existing rubrics and related documents for each learning outcome and incorporated additional feedback from faculty*" [59]. A specific criterion of the rubric, namely 'Evaluation of Different Ethical Perspectives/Concepts' was considered for rating all students' justifications (see **Table 5**). These were 1465 in total, considering the three first phases of the activity, and that the students had to justify the hierarchical ordering of six actions.

Each of the students' justifications was analyzed and scored by one of the researchers. Later, another researcher assigned scores and the differences were discussed. Only in four cases out of 717 registered justifications it was necessary to make an adjustment to the assigned score.

Students who were assigned a score of 1 to their justifications (46%), normally used the same measure as justification or stated comments as 'it was the most important'. Students who were assigned a score of 2 (40%), were able to relate at least one variable or different perspective as part of the justification but did not explore

Level	Description
Benchmark (1)	Student states a position but cannot state the objections to and assumptions and limitations of the different perspectives/concepts.
Milestone A (2)	Student states a position and can state the objections to, assumptions and implications of different ethical perspectives/concepts but does not respond to them (and ultimately objections, assumptions, and implications are compartmentalized by student and do not affect student's position.)
Milestone B (3)	Student states a position and can state the objections to, assumptions and implications of, and respond to the objections to, assumptions and implications of different ethical perspectives/concepts, but the student's response is inadequate.
Capstone (4)	Student states a position and can note the objections to, assumptions and implications of and can reasonably defend against the objections to, assumptions and implications of different ethical perspectives/concepts, and the student's defense is adequate and effective.

Table 5.
Ethical Reasoning Value Rubric (AAC&U) criteria used in classification of students' justifications, according to [59].

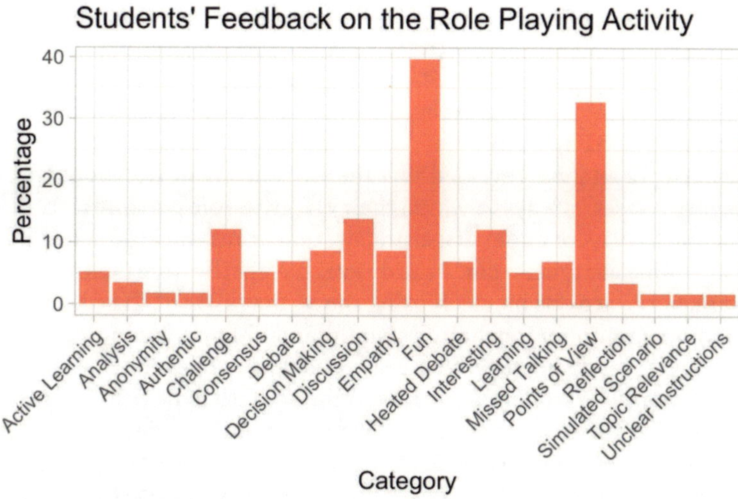

Figure 7.
Theme categories found in students' feedback.

further possible implications in their statements. The students who had a score of 3 (5%), managed to incorporate different perspectives, but their base assumption was inadequate, so the justification lost sense. Lastly, 9% of the students reached the maximum score. This result was expected, because the students are in the first year of their studies, and the ethical discrimination competence is developed throughout the duration of the entire program. Those students who achieved the maximum score, probably had a previous development of the competence, because this was the first activity of this type that was developed in the subject.

7.1 Students' feedback on the role-playing learning activity

At the end of each session in the trial, the students were asked to give their opinion on the activity through a short comment. The collection of this information was carried out through menti.com. In total, 58 comments from students were collected. In sum, 20 different themes were identified in the responses through an inductive process. **Figure** 7 shows the percentage of responses found involving each of the themes. The dominant themes were that the activity was fun (39.7%), interesting (12.1%), and that it allowed the students to share different points of view (32.8%) through discussions (13.8%) and debate (6.9%). In some cases, the students declared having engaged in heated debates and arguments with their peers (6.9%). Many highlighted the challenge of reaching a consensus (12.1%), given the different views that the adoption of the different roles led to. On the other hand, some of the students stated that the activity required them to think empathically (8.6%) with regard to the implications of their decisions in the lives of people affected by the pandemic, as well as the decisions made by other roles.

8. Conclusions

In this chapter we presented EthicApp-RP, a social platform aimed at fostering ethical reasoning, discussion and argumentation in higher education students, through a role-playing learning activity based on the jigsaw pattern. The results of the pilot activity show that the instructional design can be well enacted with business students in a synchronous online setting, that technical (see section 6.1) and pedagogical usability (see section 6.2) are positively regarded by the students, and the pedagogical goals of the activity were effectively fulfilled. The latter included eliciting students' reflection, argumentation, ethical discernments and moral reasoning through role-playing, while providing all students with equal opportunities for participation.

Sankey network analysis revealed that the students made decisions throughout the activity phases in ways in which their effective role personification was achieved, while they were doing discernment, reflection and argumentation processes by online chat messages. In addition, the justifications with which the students ranked the actions hierarchically, shows that 40% of the cohort was able to give an elaborate argument for their decisions, and 9% provided justifications of an outstanding level, which is a positive result considering the cohort was composed of business freshmen. The students considered the activity to be fun, interesting and that it fostered discussion and sharing of different points of view.

Compared to other role-playing learning activities, EthicApp-RP requires minimal logistical preparation, as the tool transparently guides students through the process, facilitating information sharing and synchronous communication among them. Furthermore, the teacher can follow the development of the activity using a progress dashboard in real time that EthicApp-RP has as a functionality. Requirements for case specification remain similar to other role-playing learning activities reported in the literature. These include the elaboration of a case based on a real or fictious situation comprising one or more ethical dilemmas, and the definition of several decision-making roles with different interests and priorities. EthicApp-RP's requirements and features make it applicable to a wide variety of learning domains and contexts in higher education, including disciplines in both science and the humanities.

Acknowledgements

This research was funded by 'Concurso de Proyectos de Investigación – FEN 2020', and CONICYT/FONDECYT Initiation into Research grant 11160211.

Author details

Claudio Alvarez[1,2*], Gustavo Zurita[3], Beatriz Hasbún[4], Sergio Peñafiel[5] and Álvaro Pezoa[6]

1 Universidad de los Andes, Chile, Facultad de Ingeniería y Ciencias Aplicadas

2 Universidad de los Andes, Chile, Centro de Investigación en Educación y Aprendizaje

3 Departamento de Control de Gestión y Sistemas de Información, Facultad de Economía y Negocios, Universidad de Chile, Santiago, Chile

4 Centro de Enseñanza y Aprendizaje, Facultad de Economía y Negocios, Universidad de Chile, Santiago, Chile

5 Departamento de Ciencias de la Computación, Universidad de Chile, Santiago, Chile

6 Universidad de los Andes, Chile, Facultad de Filosofía y Humanidades

*Address all correspondence to: calvarez@uandes.cl

References

[1] Singer P. Practical ethics: Cambridge university press; 2011.

[2] EMOL. UC suspende hasta por un año a alumnos que participaron en copia masiva por WhatsApp Chile2016 [Available from: https://www.emol.com/noticias/Nacional/2016/06/23/809292/UC-suspende-hasta-por-un-ano-a-alumnos-que-participaron-en-copia-masiva-por-WhatsApp.html.

[3] Leighton P. El alto costo de las conductas académicas deshonestas Chile2018 [Available from: http://www.economiaynegocios.cl/noticias/noticias.asp?id=452080.

[4] Vallejos RP. Experto en plagio: "el copy-paste es una realidad cultural" Chile2018 [Available from: http://lanacion.cl/2018/02/05/experto-en-plagio-el-copy-paste-es-una-realidad-cultural/.

[5] Bellé N, Cantarelli P. What causes unethical behavior? A meta-analysis to set an agenda for public administration research. Public Administration Review. 2017;77(3):327-339.

[6] Patel P. Engineers, Ethics, and the VW Scandal. IEEE Spectrum. 2015;25.

[7] Fuentes C. Chile: lindo país (coludido) con vista al mar.: CIPER; 2019 [Available from: https://ciperchile.cl/2019/03/06/chile-lindo-pais-coludido-con-vista-al-mar/.

[8] Zunger J. Computer science faces an ethics crisis. The Cambridge Analytica scandal proves it. Boston Globe. 2018;22.

[9] Transparency I. 25 Corruption scandals that shook the World: Transparency International. ; 2019 [Available from: https://www.transparency.org/news/feature/25_corruption_scandals.

[10] Sarangi S, Sharma P. Artificial Intelligence: Evolution, Ethics and Public Policy: Routledge India; 2018.

[11] Thiel CE, Connelly S, Harkrider L, Devenport LD, Bagdasarov Z, Johnson JF, et al. Case-based knowledge and ethics education: Improving learning and transfer through emotionally rich cases. Science and engineering ethics. 2013;19(1):265-286.

[12] Rafinda A, Gal T, Purwaningtyas P. Business Ethics Course On Student Moral Reasoning. Oradea Journal of Business and Economics. 2019;4(Special):60-8.

[13] Felton EL, Sims RR. Teaching business ethics: Targeted outputs. Journal of Business Ethics. 2005;60(4):377-391.

[14] Sunderland ME. Using student engagement to relocate ethics to the core of the engineering curriculum. Science and engineering ethics. 2013:1-18.

[15] Holsapple MA, Carpenter DD, Sutkus JA, Finelli CJ, Harding TS. Framing faculty and student discrepancies in engineering ethics education delivery. Journal of Engineering Education. 2012;101(2):169-186.

[16] Apostolou B, Dull RB, Schleifer LL. A framework for the pedagogy of accounting ethics. Accounting Education. 2013;22(1):1-17.

[17] Savage A, Norman CS, Lancaster KA. Using a movie to study the COSO internal control framework: An instructional case. Journal of Information Systems. 2008;22(1):63-76.

[18] Thoma SJ. Research on the defining issues test. Handbook of moral development: Psychology Press; 2006. p. 85-110.

[19] Roybark HM. Educational interventions for teaching the new

auditor independence rules. Journal of Accounting Education. 2008;26(1):1-29.

[20] Hunt SD, Vitell SJ. The general theory of marketing ethics: A revision and three questions. Journal of macromarketing. 2006;26(2):143-153.

[21] Doyle E, Frecknall-Hughes J, Summers B. Research methods in taxation ethics: Developing the Defining Issues Test (DIT) for a tax-specific scenario. Journal of Business Ethics. 2009;88(1):35-52.

[22] Vallor S. Technology and the Virtues: a Response to My Critics. Springer; 2018.

[23] Wachter-Boettcher S. Technically wrong: Sexist apps, biased algorithms, and other threats of toxic tech: WW Norton & Company; 2017.

[24] Maloney C. Weapons of Math Destruction: How Big Data Increases Inequality and Threatens Democracy. Journal of Markets & Morality. 2017; 20(1):194-197.

[25] Rosenberg M, Confessore N, Cadwalladr C. How Trump consultants exploited the Facebook data of millions. The New York Times. 2018;17:2018.

[26] ABET. Criteria for accrediting computing programs. ABET 2017 [Available from: https://www.abet.org/accreditation/accreditation-criteria/criteria-for-accrediting-computing-programs-2018-2019/.

[27] Tucker AB. Computing Curricula 1991. Commun ACM. 1991;34(6):68-84.

[28] AACSB. Ethics Education in Business Schools Report of the Ethics Education Task Force to AACSB International's Board of Directors: AACSB International; 2004 [Available from: https://www.aacsb.edu/~/media/AACSB/Publications/research-reports/ethics-education.ashx.

[29] Kohlberg L, Hersh RH. Moral development: A review of the theory. Theory into practice. 1977;16(2):53-59.

[30] Pfeffer J, Fong CT. The business school 'business': Some lessons from the US experience. Journal of management studies. 2004;41(8):1501-1520.

[31] Johnson JF, Bagdasarov Z, Connelly S, Harkrider L, Devenport LD, Mumford MD, et al. Case-based ethics education: The impact of cause complexity and outcome favorability on ethicality. Journal of Empirical Research on Human Research Ethics. 2012;7(3):63-77.

[32] Alvarez C, Zurita G, Baloian N. Applying the concept of implicit HCI to a groupware environment for teaching ethics. Personal and Ubiquitous Computing.1-19.

[33] Alvarez C, Zurita G, Baloian N, Jerez O, Peñafiel S, editors. A CSCL Script for Supporting Moral Reasoning in the Ethics Classroom. International Conference on Collaboration and Technology; 2019: Springer.

[34] Rich KL. Introduction to ethics. Nursing ethics: Across the curriculum and into practice. 2016:3-30.

[35] Cornelius N, Wallace J, Tassabehji R. An Analysis of Corporate Social Responsibility, Corporate Identity and Ethics Teaching in Business Schools. Journal of Business Ethics. 2007:117-135.

[36] Bangor A, Kortum PT, Miller JT. An empirical evaluation of the system usability scale. Intl Journal of Human–Computer Interaction. 2008;24(6): 574-594.

[37] Mulhearn TJ, Steele LM, Watts LL, Medeiros KE, Mumford MD, Connelly S. Review of instructional approaches in ethics education. Science and Engineering Ethics. 2017;23(3):883-912.

[38] Stenmark CK, Antes AL, Wang X, Caughron JJ, Thiel CE, Mumford MD. Strategies in forecasting outcomes in ethical decision-making: Identifying and analyzing the causes of the problem. Ethics & behavior. 2010;20(2):110-127.

[39] Ellertson CF, Ingerson M-C, Williams RN. Behavioral ethics: A critique and a proposal. Journal of Business Ethics. 2016;138(1):145-159.

[40] Illingworth S. Approaches to ethics in higher education: teaching ethics across the curriculum. 2004.

[41] Hess JL, Strobel J, Brightman AO. The development of empathic perspective-taking in an engineering ethics course. Journal of Engineering Education. 2017;106(4):534-563.

[42] Harris AL, Lang M, Yates D, Kruck SE. Incorporating Ethics and Social Responsibility in IS Education. Journal of Information Systems Education. 2011;22(3):183-189.

[43] Brown KM. Using role play to integrate ethics into the business curriculum a financial management example. Journal of Business Ethics. 1994;13(2):105-110.

[44] Glass R. Integrating ethics into Information systems Courses: A Multi-method Approach based on Role playing. Journal of Information Systems Education. 2020;6(4):2.

[45] Génova G, González MR. Teaching ethics to engineers: A socratic experience. Science and engineering ethics. 2016;22(2):567-580.

[46] Prince RH. Teaching engineering ethics using role-playing in a culturally diverse student group. Science and Engineering Ethics. 2006;12(2):321-326.

[47] El'fred H, Koh HC. The influence of organizational and code-supporting variables on the effectiveness of a code of ethics. Teaching Business Ethics. 2001;5(4):357-373.

[48] Mintz SM, Morris RE. Ethical obligations and decision making in accounting. Boston, et al: McGraw-Hill Irwin. 2008:56-57.

[49] Hess JL, Fore G. A systematic literature review of US engineering ethics interventions. Science and engineering ethics. 2018;24(2):551-583.

[50] Thistlethwaite JE, Davies D, Ekeocha S, Kidd JM, MacDougall C, Matthews P, et al. The effectiveness of case-based learning in health professional education. A BEME systematic review: BEME Guide No. 23. Medical teacher. 2012;34(6):e421-ee44.

[51] Jerez O, Andrade C, Aranda R. Planificaciones generales para trabajo en aula con Dilemas Éticos: Facultad de Economía y Negocios, Universidad de Chile.; 2019.

[52] Arredondo F. La experiencia de la Metodología de Casos aplicado a la Enseñanza Ética200. Available from: http://sitios.itesm.mx/va/diie/tecnicasdidacticas/documentos/2LaExperienciaEtica.pdf.

[53] Kolodner JL, Guzdial M. Theory and practice of case-based learning aids. Theoretical foundations of learning environments. 2000:215-242.

[54] Kim S, Phillips WR, Pinsky L, Brock D, Phillips K, Keary J. A conceptual framework for developing teaching cases: a review and synthesis of the literature across disciplines. Medical education. 2006;40(9):867-876.

[55] Nokelainen P. An empirical assessment of pedagogical usability criteria for digital learning material with elementary school students. Journal of Educational Technology & Society. 2006;9(2):178-197.

[56] Hernández Leo D, Asensio-Pérez JI, Dimitriadis Y. Computational representation of collaborative learning flow patterns using IMS learning design. Journal of Educational Technology & Society 2005; 8 (4): 75-89. 2005.

[57] Zurita G, Hasbun B, Baloian N, Jerez O. A blended learning environment for enhancing meaningful learning using 21st century skills. Emerging issues in smart learning: Springer; 2015. p. 1-8.

[58] Zurita G, Baloian N, Peñafiel S, Jerez O, editors. Applying Pedagogical Usability for Designing a Mobile Learning Application that Support Reading Comprehension. Multidisciplinary Digital Publishing Institute Proceedings; 2019.

[59] Association of American Colleges and Universities. Retrieved July 15. Ethical reasoning VALUE rubric. 2010 [Available from: https://assessment.provost.wisc.edu/wp-content/uploads/sites/92/2017/02/EthicalReasoning.pdf.

Chapter 8

The Rise of Virtual Reality in Online Courses: Ethical Issues and Policy Recommendations

Clement Longondjo Etambakonga

Abstract

While ethical issues related to the adoption of Virtual Reality (VR) technology is analyzed across sectors from construction, architecture, retail, engineering, healthcare, less attention is paid to the ethical concerns in online courses. Using an inductive qualitative content analysis and observation in the business schools, this chapter aims to shed light on the ethical issues that may occur as results of use of VR technology in online courses. The findings indicate that the use of VR technology in online courses raises several technical and social/ethical issues. These issues comprise concerns related to record of personal data, which could be deployed in ways that threaten personal privacy, private neglect of users' own real bodies and actual physical environments, and other moral and social security risks related with the way VR confuses the distinction between face-to-face and virtual learning. As these ethical issues raise questions about public policy, the chapter makes several recommendations that elaborate a platform for further discussion. It is argued that there is a need for a wider vision that looks beyond the teaching technological issues to those linked to students and teachers' conducts, and institution policies.

Keywords: Virtual reality technology, online courses, ethical issues, public policy

1. Introduction

In the past two decades, more particularly since the spread of Covid19 pandemic that has forced worldwide shutdown of business schools, online education has increased remarkably and has involved increasing numbers of teachers lecturing at a distance for the first time. This resulted in tremendous compulsory use of Virtual Reality (VR) technology in online courses serving as the learning platform. VR is a computer-based technology that offers students with a highly collaborative and multi-sensory learning experience [1, 2]. It delivers real-time commitment and instantaneous feedback regarding student performance and challenge [3]. However, even though there has been increasing attention devoted to ethical issues related to the adoption of Virtual Reality (VR) across sectors from construction, architecture, retail, engineering, and healthcare, there has been less attention dedicated to ethical issues with the widespread use of VR in online learning. It is argued that during the Covid19 pandemic, the number of online courses in business schools is growing significantly which have several ethical issues [2, 4]. This chapter addresses ethical issues faced by teachers and students in the business schools when using VR technology on online courses.

Using an inductive qualitative content analysis as one of the several qualitative methods currently available for analyzing data and interpreting its meaning [5], this chapter tries to answer the following ethical questions: Does the use VR technology in online courses guarantee privacy, security and confidentiality of the participants? Does a dependence on virtual reality technology in online teaching promote equity and diversity among students? What are ethical issues faced by students and teachers in their interaction using virtual reality technology in online courses? The structure of the chapter is as follows. The next section conceptualizes and defines virtual reality technology and briefly reviews various ethical issues related to the use of virtual reality technology. Section 3 discusses VR regarding online courses and the benefits of VR to online education are reviewed. Section 4 contains an ethical issues analysis faced by teachers and students and the last session discusses policy recommendations. The conclusion summarizes key ethical implications of use of VR in online courses, particularly those facing teachers and students.

2. Virtual reality technology

Virtual Reality (VR) has become a most important resource and facet of our present and future learning. The term VR was first used in 1974 Myron Krueger describing specific environment, an artificial reality display as video place [6]. Today, we are experiencing the most dauting and exhausting learning perspective settings with virtual reality. One of the most commonly uses of the term defines the virtual reality systems as systems that use head-mounted exhibitions, data gloves and data suits to replicate an immersive, highly interactive computer-generated, multi-sensory information environment, in which user becomes participant in real time. Generally, many scholars use virtual reality interchangeably with cyberspace and artificial reality. For example, Brey defines virtual reality as "a three-dimensional interactive computer-generated environment that incorporates a first-person perspective" [7, p 5]. Sandra Helsel defines virtual reality as a process that enables users to become participants in abstract spaces where the physical machine and physical viewer do not exist [8].

In this chapter, VR is considered a technology that persuades students and teachers that they are in real time that replace face-to-face through the use data established shaped by a computer. It covers the entire field, comprising artificial reality, internet, and a third person and telepresence, which according to Hilary McLellan, is the sensation of being present at a distant place from where one is truly located, with the capacity to control objects at that remote location. However, the key characteristics that emerge from the virtual reality definitions include the use of three-dimensional graphics, the interactivity and a first-person perspective [9]. While interactivity requires that the represented environment must enable for manipulation, which implies the modification of aspects of the environment in a fairly direct way, such interactivity entails three-dimensional graphics. For instance, a first-person perspective, involves that the environment is recognized and interacted with from a specific point, that indicates a degree of immersion in a world, rather than the experience of the world as an entity that can be monitored from the outside [7, 10, 11].

Even though, the immersion in imaginary space as a key element of VR is important and allows one to forget about the physical distance among participants, it however cannot exactly replace the warmest face-to-face education. Such immersion is only limited by our imagination and how we choose to build the virtual world [12]. It is worth noting that VR technology offers several benefits, particularly in education by providing a safe, helpful and conducive environment for students to improve their learning process [12, 13]; but it poses serious issues ranging from potential mental issues, personal neglect of users' own actual bodies and its technology may be used to record personal data which may threaten personal privacy and risks the manipulation of users' beliefs, emotions, and behaviors [14, 15]. As moral obligation represents a constraint which not only mitigates a virtual world's experience, but which may prove antithetical to the medium's long-range social influence [14], therefore the focus of this chapter is on the ethical issues faced by multi-users of VR systems in online courses.

3. Virtual reality and online courses

The literature in VR is used by several different scholars with many meanings. Even though, VR was mainly used to play the most recent computer games, it is today a constantly advancing new computer technology, which offers great opportunities for the education sector. In education, it is used as a path for students and teachers to envision, control and interact with computers and tremendously complicated data [2]. By simulating learners and teachers in a face-to-face physical presence in the real environment, VR is an entirely immersive, engaging and interactive experience of another reality in which the participants feel completely absorbed in the environment by means of special human-computer interface equipment [11]. VR has the possibility to be a formidable new instrument in the teaching space. It provides the best instrument for learning by submerging learners and teachers in an environment as close to face-to-face. VR creates environments in which participants generate various social and disciplinary cultures, with unique communication patterns, norms, values, and interaction systems [16, 17], as variations in communication forms change classroom-based conceptions of teaching and online learning.

The online course systems are web-based packages for supplying, stalking, and managing courses over the cyberspace. Anderson and Simpson pointed out that online course entails the application of innovations in technology to direct, model and provide the learning substance, and to facilitate two-way communication between students and teacher [2]. It is increasingly agreed that online courses are a more useful and flexible method to taking courses that will lead to a degree. Virtual learning contains elements such as chat rooms, whiteboards, discussion forums and quizzes that enable students and teachers to communicate online and share the whole course content likely to achieve the learning goals. For instance, John C. Thomas and Rory Stuart list seven roles for online education using virtual reality including, investigating existing places and things that students would not otherwise have access to, explore real things that, without changes of scale in size and time, could not otherwise be well explored, create places and things with altered qualities, interact with people who are in distant locations through global clubs with a shared interest, or collaborations on projects between students from different parts of the world, interact with real people in non-realistic ways, create and manipulate abstract conceptual representatives, like data structures and mathematical functions and interact with virtual beings, such as representations of historical figures and agents who are representatives of different philosophies and viewpoints participating in simulated negotiations [18, p. 209].

For many of benefits, business schools are resorting to VR to deliver education online. The use of VR in education offers several benefits ranging from students learning experience through computer-based interaction to the development of student's information and communication technology skills [7, 19]. It has been argued that the use of VR increases students' engagement in online learning by interacting them in multiple ways [20]. The use of VR in education will continue to increase student's level technological expertise in a world [21], in a world where all activities tend to be digitalized. However, despite several benefits provided by the use of VR in online learning, teaching and learning at a distance raises several ethical issues, which are even complex than those encountered by face-to-face teachers and students [2]. Zembylas & Vrasidas pointed out that online settings create sites that are 'supportive of hybrid identities, complex discourses, and multiple relations among learners.' [22, p.61]. The use VR systems in online education may raise several ethical issues that go beyond the nature and applications applied to technology to the burden of some social groups and unethical behaviors faced by learners and teachers.

4. Ethical issues in online teaching

The introduction of virtual reality (VR) technology in online courses has raised several ethical issues. My focus of this section is the consideration of rights of students and teachers in regard to the dangers they are exposed to in online learning environments that introduce new and/or intensify existing ethical issues as they interact. These moral issues occur in part because electronic environments allow new kinds of behaviors that are simple to perform in electronic environments, which may entail new ethical rules [23]. As VR online relies upon the internet that is wrought with potential risks, what might we have to worry about once more business schools adopt the technology in their online courses. The purpose of this chapter is not to furnish responses to all ethical issues related to online courses using VR, but to raise further inquiries for shaping solutions to the ethical issues I consider, based on Kidder's advice to resolve our ethical concerns through energetic self-reflection. Relying on an inductive qualitative content analysis of archival data and observation related to the ethical issues faced by students and teachers using VR technology in online courses, I have identified ten key types of ethical issues, which are briefly discussed below:

4.1 Privacy and confidentiality

The issue of privacy and confidentiality raises questions about what kind of data is recorded and stored? Who gives permission to store the data? The use of VR technologies in online courses may be used to record, store and even share personal data which could be used in ways that jeopardize personal privacy and present a risk linked to misuse of users' behaviors, values and emotions [2]. Privacy is essential in keeping precious conditions of ethical humanhood. O'Brolcháin et colleagues note that people, if they are to that expand themselves and explore their opinions, or again behaving in certain manners need a degree of privacy [24]. However, reduced privacy influences the development of individual ethical personalities. In VR online courses, both students and teachers will no longer have any private space to make errors and explore distinct aspects of themselves as they are immersed in a digital environment [23]. In the European Union for instance, individual rights are protected by the general data protection regulation that ensure personal data protection to guarantee privacy. It is worth noting that VR technology, in particular online learning raises new privacy issues or exacerbates existing ones [2]. However, these issues are likely to be applied in the business schools that are currently using online education, in which many of their activities are recorded electronically including staff meetings, classroom, students group projects and many other online activities.

4.2 Safety and security

The issue of safety and security raises a question about how to ensure transfer of the correct data to the right address? As any connected device that uses internet, VR technology in online teaching does not ensure essential security mechanisms for data storage and sharing and communications between VR devices, servers might be sent unencrypted. Like smart phones that can secretly collect any kind of information from everywhere, if VR earphones become universal, everyday devices, then someone might be able to track what you are watching at any time [25]. However, what occurs if someone hacks VR earphones and introduces a visual attack that could cause harmful real-world reactions? There could be several ways hackers put personal information into harm's way if needed.

4.3 Informed consent

The issue of informed consent raises a question about whether other people publish information about/pictures of me? Regarding the information gathered as teachers interact with students to build up a meaningful personal experience, the following questions could be raised for the informed consent: At what point does a teacher's interest in knowing more about a student in order to make a significant connection to learning interfere on the student's personal right to privacy? Do teachers need to seek for consent to watch students' online interactions, to review and reassess their online contributions and to 'eavesdrop' on their 'conversations'? [2]. The VR in online learning environment offers greater opportunities for teachers to know about students and to have much of that knowledge recorded always [24]. Many students who interact on online virtual environments during the courses, which becomes the norm, sometimes forget that they can be observed. But informed consent and being able to control the use of personal information is definitely a person's right [2]. When students register at a business school, there is a need to seek permission to gather and use student's personal data as part of the application process to provide the organization with numerical data about the nature of the student group.

4.4 Equality, equity and diversity

The issue of equity and diversity raise questions about whether the use of VR in online courses promote equality, equity and diversity for students? Can VR in online courses ensure education for all? Can social, cultural and academic values be successfully transmitted using VR in online learning? As VR in online courses requires internet connection and computers which limit access to many social groups (i.e., poor people) - social exclusion of people who cannot afford to buy a computer or get connected to the internet. Lack of information and communication technology equipment among low-income people, exclude them to online course systems. However, Wedemeyer stated that" Instruction should be available any place where there are students—or even only one student—whether or not there are teachers at the same place at the same time" [26, p.36]. This statement underlines the ethical aspect of equality of access for all on online learning using VR as a moral obligation. There are moral issues of culture in online course using VR, including the of imbalances arising from dominant cultural morals represented in teaching materials and methods [27], and possible miscommunication among participants in online discussions arising from cultural differences [28]. Additionally, most online courses in business schools are designed in English, non-English speakers can be excluded.

4.5 Autonomy

The issue of autonomy in online learning using VR raises questions such as are VR in the online education environment conducive for learning freedom or do they threaten to undermine it? Autonomy as self-control plays a key role in ethics as it is about to obey only yourself –to be able to deliberate and make decisions without being influenced by external sources [24]. To be autonomous, individuals will need access to appropriate information from relevant sources without constraints and to be able to choose for themselves according to their own ideas and values in order to make decisions. However, VR technology in online learning acts as a gatekeeper of information, which causes a risk to the informational condition of autonomy. The institution has control over information posted, can also control how people perceive and understand the world of learning. Additionally, VR technology in online courses raises serious concerns related to personal neglect of users' own real bodies and real physical environments [29].

4.6 Copyright and plagiarism

The issue of copyright and plagiarism raises questions such as how to protect copyrighted data, students and teachers' contributions and materials used for teaching, from being exchanged illegally? It is increasingly known that the prohibited copying of copyrighted media (e.g. texts, music works, movies and software programs) is prevalent throughout in education [2]. Additionally, many students and teachers who engage in such misconduct practices do not think themselves to be doing something that is obviously morally wrong. This is certainly true for college students. As Glass and Wood have reported that a large majority of students do not recognize the illegal copying of software as unethical [30]. Moreover, plagiarism is widespread in many business schools, where it is one of the biggest forms of academic dishonesty. Copying the ideas or work of another person without citing the source, including books, extracts of articles, tables, diagrams and material from the internet or other electronic sources is common among students. Uunauthorized or inappropriate use of computers, calculators and other forms of technology in coursework, assignments, Brey reported that assignments handed in by students may turn out to be copied from other students or to be taken over, in part or in whole, from existing published works [23]. VR in online courses tools such as computers and the internet only add to the way that students have at their possession to perpetrate plagiarism.

4.7 Ownership of data

The issue of ownership of data raises the question about who owns the data published on social networking sites? Does a teacher or a student keeps the ownership of his own data, or he/she loses it in the moment he/she accepts the business school's terms and conditions? Are foundational rights of individual subjects recognized in VR technology in the online learning environment, and what it takes to protect them against obstructions? These few questions among many others arise against the background of disputed relations of ownership between an owner and his/her property. VR technology operates through gathering and processing of huge data. Ownership of data, access and control are critical moral issues in the sharing of VR in online courses data [31]. In many business school policies, data generated and posted to VR learning environments are owned by the institutions.

4.8 Online bullying and hacking

Online bullying and hacking raise the question about how individuals can be protected online against personal attacks and stalking? Billions of students are bullied everyday worldwide, as many students think that bulling is a tradition of passage in a student's life. This includes harassment and intimidation that takes place online using pictures or words and is difficult to control. It takes several forms of exclusion, threats, aggression to public humiliation mainly among the students at the universities and business schools, these are morally wrong. VR in online courses could be used as a platform for online bullying or harassment, racists could also use it as platform for bullying behaviors in the shadow of anonymity. Additionally, hacking involves interfering on someone's personal computer through distant access, purposely changing files to which one has not been granted access, liberating computer viruses, stealing passwords or files, exposing personal information, and stealing electronic money [32]. Teachers and students at VR in online courses may get involved in hacking for a range of reasons - they may just be unaware that they are breaking into a computer system, they may just be curious, they may be out to harm someone, they may want to benefit themselves, or they may have entirely different reasons [23].

4.9 Control and surveillance

The issue of control and surveillance raises question about how to control the exchange between students and teachers or among students in virtual environments? How to ensure surveillance of students in online courses? In one side, virtual technology allows anonymous communications because of the use of internet from harassment to fraud which are difficult to track and appear difficult to solve. On the other side, surveillance of students in all online courses is an issue. Actually, in online courses using virtual technology, teachers are to track students' participation, the number of responses they post and what they read, when they read it. Virtual courses provide teacher with a permanent record, which are often used to give us information about a student's performance, however, such surveillance is hidden, and concern and action on the part of those being observed is driven by uncertainty [24].

4.10 Freedom of speech versus hate messages

The issue of freedom of speech versus hate speech raises questions about what are the limits of what can be published? Who decides what is acceptable? Even though, in most countries some degree of freedom of speech is ensured in the constitution, but there are limitations for hate speech, defamation, and obscenity. In business schools, there may be various electronic ways of exchanging messages between students and teacher or among students and/or with external people to the organization, which are important collaborative virtual settings. As in face-to-face interaction, these VR forms of interaction can be used to send threatening, obscene, provocative or harassing messages [24]. These may include discriminatory, sexist, or/and racist messages, used to tease fellow students or teachers based on their socio-cultural affiliations. However, such messages are usually not considered to be acceptable in an academic setting, as educators strive to ensure that the classroom functions as a safe, non-hostile environment for students and teachers in virtual courses as do face-to-face classes [24]. **Table 1** below summarizes ethical issues related to the use of VR technology in online courses and the questions they raised

Ethical issues	Questions they raised
Privacy and confidentiality	No access of third parties. What kind of data is stored? Who gives permission to store/retrieve the data?
Safety/ security	How to ensure transfer of the correct data to the right address?
Informed Consent	Can other people publish information about/pictures of me?
Equality, equity and diversity	The equity and diversity raise questions such as does the use of VR in online courses promote equality, equity and diversity for students? Can VR in online courses ensure education for all? Can social, cultural and academic values be successfully transmitted using VR in online learning?
Copyright and plagiarism	How to protect copyrighted data, students and teachers' contributions and materials used for teaching e.g. music and movies, from being exchanged illegally?
Autonomy	Are VR environments in online education conducive for learning freedom or do they threaten to undermine it?
Ownership of data	Who owns the data published on social networking sites?
Online bullying and hacking	How can individuals be protected online against personal attacks and stalking?
Control and surveillance	how to control the exchange between students and teachers or among students in virtual environments?
Freedom of speech versus hate speech	What are the limits of what can be published? Who decides what is acceptable?

Table 1.
Summary of Ethical Issues and Questions they raised.

5. Policy recommendations

Given the limited number of ethical issues discussed in this chapter in regard to the increasing moral issues related to the use of VR technology in online learning, the following set of recommendations provided here serve as a regulating starting point, a framework for future debates:

- There is a necessity to increase communication and discussion among information and communication technology professionals, businesses, business schools and governments to get things right and address moral issues around VR technology in online learning.

- For instance, in addressing security issues, the business schools should investigate the track record of the producer and the devices to protect against the hackers.

- As there are many new VR technologies, software and devices for online learning, universities, students/teachers' behavior, and the regulation should need to change to keep up with any new changes of VR in online learning.

- Business schools should have policies that address all identified ethical issues to ensure safer, an inclusive and responsible learning environment.

- A VR in online education trainings should be run to all users including teachers and students to increase their technological skills such as the tools to support activities, the key tasks to accomplish and the learning environment, these skills need to be further assessed before using VR technology.

- There is a need for a qualified and available technical team with appropriate resources needed to respond to the needs of teachers and students during online courses using VR.

- VR technology users should be aware of all ethical issues related the use of VR in online courses and the potential solutions on how to address them.

- VR technology users should be aware that as compared to the watching of traditional movies, the effect of VR immersion settings and the related danger of users suffering mental health trauma will gradually increase.

- VR technology users should be aware of new threats concerning surveillance and data protection during the use of VR in online courses.

6. Conclusion

Since the spread of the Covid19 pandemic, teaching students in distance has become a must for many business schools. There are many ethical issues related to the use of VR technology, particularly in online learning; however, this chapter has considered and discussed only ten of them. The study suggests that, even though online courses using VR technology offer several benefits, they pose several social and technical issues. Some of the issues that have been discussed here range from privacy and security of the users to the social/ethical issues, and recommendations have been made as a starting point to stimulate future discussion. It is argued that there is a need for a wider vision that looks beyond the teaching technological issues to those linked to students and teachers' conducts and institutions' policies. Further research is needed to understand the extent of the quality of learning, class interactions, social and ethical values in VR online learning as compared to face-to- face educational settings. Recommendations for future discussion have been provided above.

Author details

Clement Longondjo Etambakonga
Management and Organization Department, Rennes School of Business, France

*Address all correspondence to: clement.longondjo@rennes-sb.com

References

[1] Madary M, Metzinger T. Real Virtuality: A Code of Ethical Conduct. Recommendations for Good Scientific Practice and the Consumers of VR-Technology. Frontiers Robotics AI 2016; 3, 3.

[2] Anderson B, Simpson M. Ethical issues in online education. *Open Learning: The Journal of Open, Distance and E-Learning* 2007; *22*(2), 129-138.

[3] Stewart B, Hutchins H M, Ezell S, De Martino D, Bobba A. Mitigating challenges of using virtual reality in online courses: a case study. *Innovations in Education and Teaching International 2010*; *47*(1), 103-113.

[4] Collins D, Weber J, Zambrano R. Teaching business ethics online: Perspectives on course design, delivery, student engagement, and assessment. *Journal of Business Ethics* 2014; *125*(3), 513-529.

[5] Schreier M. Qualitative content analysis in practice. Thousand Oaks, CA: Sage; 2012.

[6] Krueger M W. Artificial Reality II. Reading, MA: Addison-Wesley; 1991.

[7] Brey P. The ethics of representation and action in virtual reality. *Ethics and Information technology* 1999; 1(1) 5-14.

[8] Hclsel S. Virtual Reality and Education. *Educational Technology* May 1992; 38-42.

[9] McLellan H. Situated learning: Multiple perspectives. *Situated learning perspectives* 1996; 5-17.

[10] Steuer J. Defining virtual reality: Dimensions determining telepresence. *Journal of communication* 1992; *42*(4), 73-93.

[11] Halarnkar P, Shah S, Shah H, Shah H, Shah A. A review on virtual reality. *International Journal of Computer Science Issues (IJCSI)* 2012; *9*(6), 325.

[12] Hussein M, Nätterdal C. The benefits of virtual reality in education-A comparison Study; 2015.

[13] Carvalho Mano R M. *The benefits of Virtual Reality in Education* (Doctoral dissertation, Hochschule für angewandte Wissenschaften Hamburg); 2019.

[14] Cranford M. The social trajectory of virtual reality: Substantive ethics in a world without constraints. *Technology in Society* 1996; *18*(1), 79-92.

[15] Kenwright B. Virtual Reality: Ethical Challenges and Dangers [Opinion]. *IEEE Technology and Society Magazine 2018; 37*(4), 20-25.

[16] Whittaker S. Things to talk about when talking about things. *Human Computer Interaction 2003;* 18(2), 149-170.

[17] Bakardjieva M, Feenberg A. Involving the virtual subject. *Ethics and Information Technology 2000*; 2, 233-240.

[18] Thomas J C, Stuart R. Virtual reality and human factors. In *Proceedings of the Human Factors Society Annual Meeting* 1992; 36 (3) 207-210.

[19] Satterfield D, Kelle S. Ethical issues in online education. In *Advances in The Human Side of Service Engineering* 2017; 257-266.

[20] Swartz L B, Cole M T. Students' Perception of Academic Integrity in Online Business Education Courses. *Journal of Business and Educational Leadership 2013*; *4*(1), 102.

[21] Waight CL, Stewart BL. Valuing the adult learner in e-learning: A conceptual model for corporate settings. *Journal of Workplace Learning* 2005 17, 337-345.

[22] Zembylas M, Vrasidas C. Levinas and the 'inter-face': the ethical challenge of online education, *Educational Theory* 2005; 55(1), 61-78.

[23] Brey P. Social and ethical dimensions of computer-mediated education, *Journal of Information, Communication and Ethics in Society* 2006; 4(2), 91-101.

[24] O'Brolcháin F, Jacquemard T, Monaghan D, O'Connor N, Novitzky P, Gordijn B. The convergence of virtual reality and social networks: threats to privacy and autonomy. *Science and engineering ethics 2016*; *22*(1), 1-29.

[25] Tsia P. The Rise of VR and Its Impending Security Risks 2016; available at: https://www.tripwire.com/state-of-security/security-data-protection/rise-vr-impending-security-risks/, accessed on 24th December 2020.

[26] Wedemeyer C A. Learning at the back door: reflections on non-traditional learning in the lifespan Madison, WI, University of Wisconsin; 1981.

[27] Pincas A. Culture, cognition and communication in global education, *Distance Education 2001;* 22(2), 30-51.

[28] Goodfellow R, Hewling A. Reconceptualising culture in virtual learning environments: from an 'essentialist' to a 'negotiated' perspective, E–Learning 2005; 2(4), 355-367.

[29] Spiegel J S. The ethics of virtual reality technology: social hazards and public policy recommendations. *Science and engineering ethics* 2018; *24*(5), 1537-1550.

[30] Glass R, Wood W. Situational Determinants of Software Piracy: An Equity Theory Perspective. *Journal of Business Ethics 1996;* 15: 1189-1198.

[31] Hummel P, Braun M, Dabrock P. Own data? Ethical reflections on data ownership. *Philosophy & Technology* 2020; 1-28.

[32] Forester T, Morrison P. Computer Ethics: Cautionary tales and ethical dilemmas in computing, 2nd ed. Cambridge and London: MIT Press; 1994.